Standards Practice Book

For Home or School

Grade K

Houghton
Mifflin
Harcourt

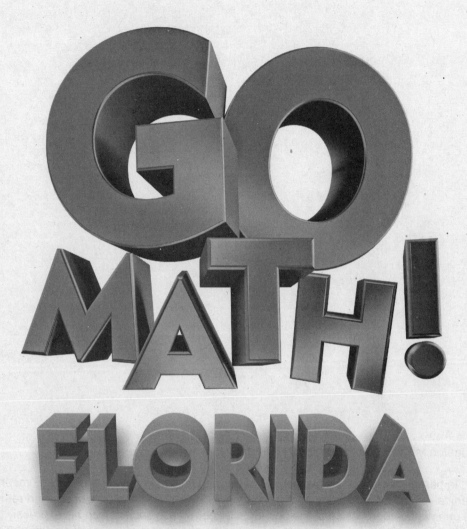

INCLUDES:

- Home or School Practice
- Lesson Practice and Test Preparation
- English and Spanish School-Home Letters
- Getting Ready for Grade 1 Lessons

Number and Operations

Representing, relating, and operating on whole numbers, initially with sets of objects

1 Represent, Count, Read, and Write Numbers 0 to 5

Domains Counting and Cardinality

2 Compare Numbers to 5

Domain Counting and Cardinality

5 Addition

Domains Operations and Algebraic Thinking

6 Subtraction

Domains Operations and Algebraic Thinking

Geometry and Positions

Describing shapes and space

9 Identify and Describe Two-Dimensional Shapes

Domains Geometry

10 Identify and Describe Three-Dimensional Shapes

Domain Geometry

Measurement and Data

Representing, relating, and operating on whole numbers, initially with sets of objects

11 Measurement

Domains Measurement and Data

12 Classify and Sort Data

Domains Measurement and Data

End-of-Year Resources

Getting Ready for Grade 1

These lessons review important skills and prepare you for Grade 1.

Table of Contents
Florida Lessons

School-Home Letter

Dear Family,

My class started Chapter 1 this week. In this chapter, I will show, count, and write numbers 0 to 5.

Love, _____

Vocabulary

one a number for a single object

two one more than one

Home Activity

Use this five frame and counters, such as buttons. Have your child place counters in the five frame to show the numbers 0 to 5. For 0, have your child place one counter in the five frame, and then remove it. Together, practice writing the numbers 0 to 5.

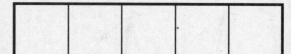

Literature

Look for this book in a library. This book will reinforce your child's counting skills.

Fish Eyes: A Book You Can Count On by Lois Ehlert. Voyager Books, 1992.

Carta
para la casa

Querida familia:

Mi clase comenzó el Capítulo 1 esta semana. En este capítulo mostraré, contaré y escribiré números del 0 al 5.

Con cariño, _____

Vocabulario

uno el número de un solo objeto

dos uno más que uno

Actividad para la casa

Use este cuadro de cinco y fichas, tales como botones. Pídale a su hijo que ponga las fichas en el cuadro para mostrar los números del 0 al 5. Para 0, pídale que coloque una ficha en el cuadro de cinco y luego que la quite. Juntos, practiquen la escritura de los números del 0 al 5.

Literatura

Busque este libro en una biblioteca. Este libro ayudarán a su hijo a reforzar la destreza de contar.

Fish Eyes: A Book You Can Count On
por Lois Ehlert.
Voyager Books, 1992.

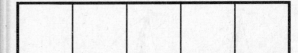

School-Home Letter

Dear Family,

My class started Chapter 2 this week. In this chapter, I will learn how to build and compare sets to help me compare numbers.

Love, _____

Vocabulary

same number

There are the same number of circles and triangles.

greater

The number of circles is greater than the number of triangles.

less

The number of circles is less than the number of triangles.

Home Activity

Gather two sets of five household items. Line some of them up on a table in two groups of different quantities. Ask your child to count and tell you how many are in each set. Have your child point to the set that has the greater number of objects. Then ask your child to point to the set with the number of objects that is less.

Change the number in each group and repeat the activity.

Literature

Look for this book in the library. It will help reinforce concepts of comparing.

More, Fewer, Less
by Tana Hoban.
Greenwillow Books, 1998.

Carta
para la casa

Querida familia:

Mi clase comenzó el Capítulo 2 esta semana. En este capítulo, aprenderé cómo construir y comparar conjuntos que me ayuden a comparar números.

Con cariño, _____

Vocabulario

igual número

Hay igual número de círculos y triángulos.

mayor

El número de círculos es mayor que el número de triángulos.

menor

El número de círculos es menor que el número de triángulos.

Actividad para la casa

Reúna dos conjuntos con cinco elementos de la casa. Alinee sobre la mesa algunos de ellos en dos grupos de diferentes cantidades. Pídale a su hijo que cuente y diga cuántos hay en cada conjunto. Dígale que señale el conjunto que tiene el mayor número de objetos. Luego, pídale que señale el conjunto con el menor número de objetos.

Cambie el número en cada grupo y repita la actividad.

Literatura

Busque este libro en la biblioteca. Este libro ayudará a su hijo a reforzar los conceptos de más y menos.

More, Fewer, Less por Tana Hoban. Greenwillow Books, 1998.

Name _____

Same Number

_____

- - - - - - -

_____

_____

- - - - - - -

_____

DIRECTIONS I. Compare the sets of objects. Is the number of dolphins greater than, less than, or the same as the number of turtles? Count how many dolphins. Write the number. Count how many turtles. Write the number. Tell a friend what you know about the number of objects in each set.

Lesson Check

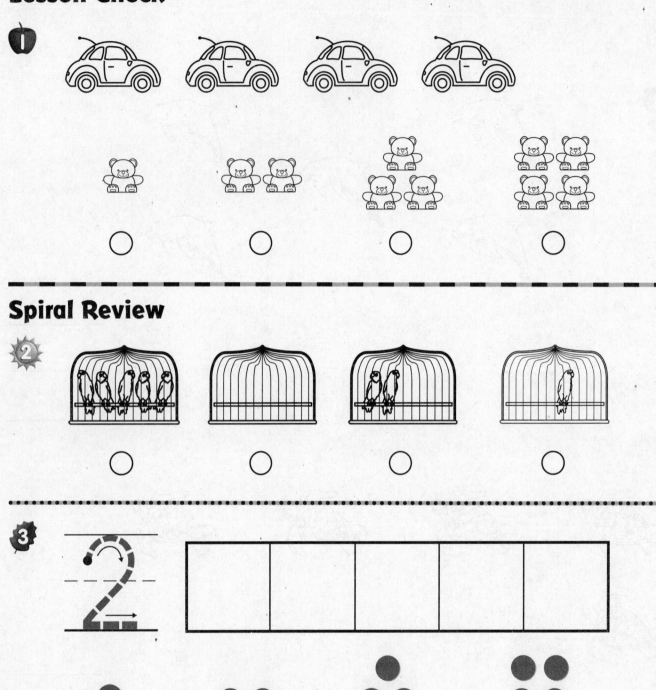

Spiral Review

DIRECTIONS **1.** Which set shows the number of bears is the same as the number of cars? Mark under your answer. **(Lesson 2.1)** **2.** Which bird cage has 0 birds? Mark under your answer. **(Lesson 1.9)** **3.** Trace the number. How many counters would you place in the five frame to show the number? Mark under your answer. **(Lesson 1.1)**

Name _____

Greater Than

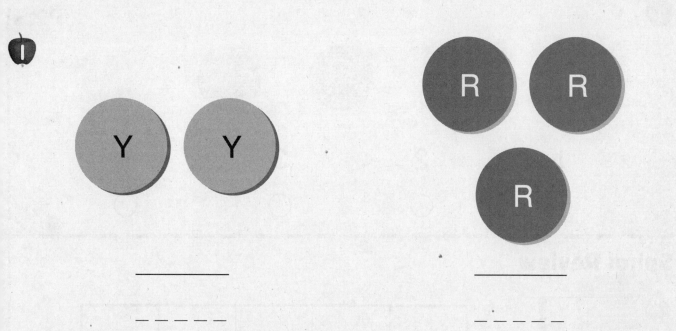

_____   _____

- - - - -   - - - - -

_____   _____

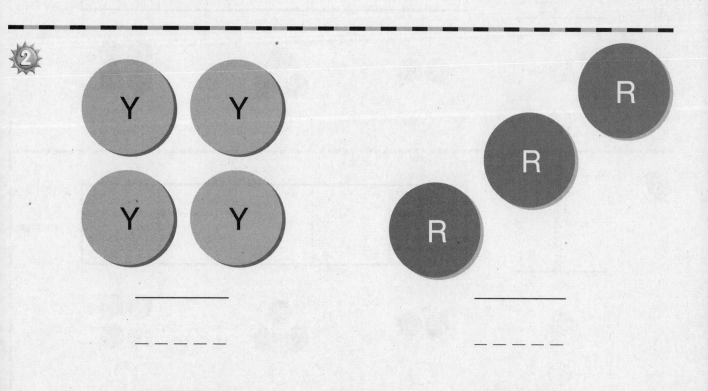

_____   _____

- - - - -   - - - - -

_____   _____

DIRECTIONS 1–2. Place counters as shown. Y is for yellow, and R is for red. Count and tell how many are in each set. Write the numbers. Compare the numbers. Circle the number that is greater.

Lesson Check

1

1   2   3   4

○   ○   ○   ○

Spiral Review

2

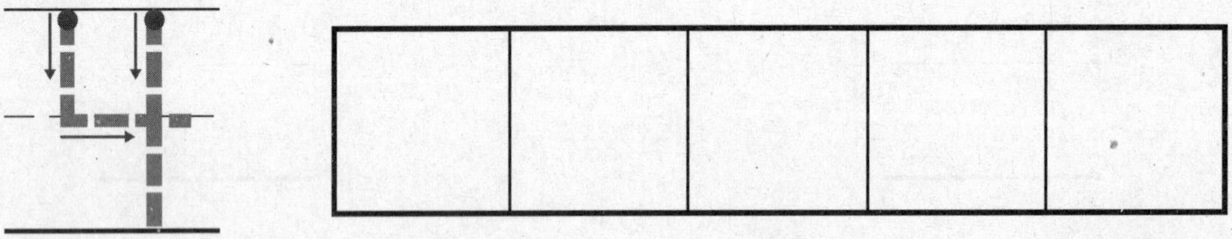

○   ○   ○   ○

3

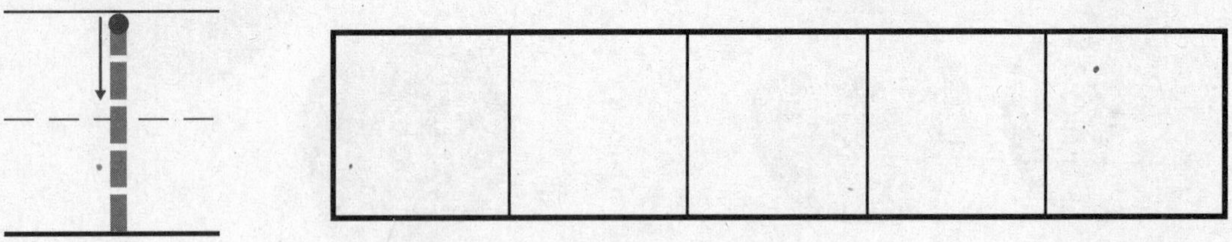

○   ○   ○   ○

DIRECTIONS   **1.** Mark under the number that is greater than the number of counters. **(Lesson 2.2)**   **2–3.** Trace the number. How many counters would you place in the five frame to show the number? Mark under your answer. **(Lessons 1.3, 1.1)**

Name _____

Less Than

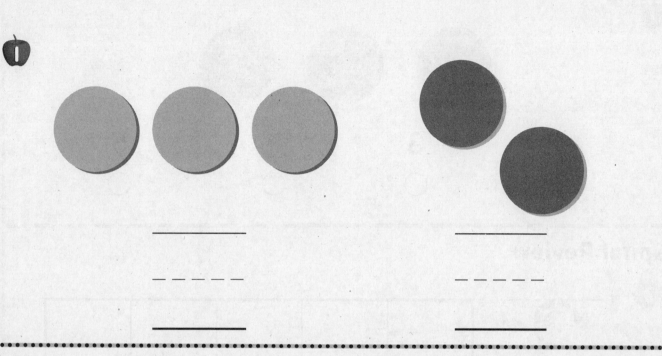

_____

- - - - -

_____

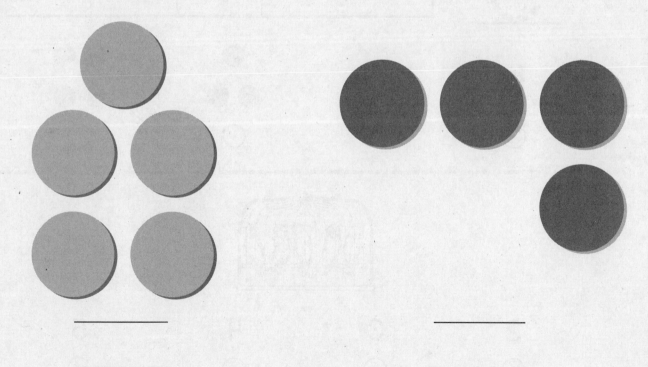

_____

- - - - -

_____

DIRECTIONS 1–2. Count and tell how many are in each set. Write the numbers. Compare the numbers. Circle the number that is less.

Chapter 2

Lesson Check

1

2	3	4	5
○	○	○	○

Spiral Review

2

●	●●	● ●●	●● ●●
○	○	○	○

3

2	3	4	5
○	○	○	○

DIRECTIONS   1. Mark under the number that is less than the number of counters. **(Lesson 2.3)   2.** Trace the number. How many counters would you place in the five frame to show the number? Mark under your answer. **(Lesson 1.3)   3.** Count how many birds. Mark under your answer. **(Lesson 1.6)**

P32   thirty-two

Problem Solving • Compare by Matching Sets to 5

 1

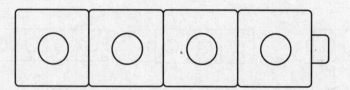

_____

- - - - - - - - -

_____

_____

 2

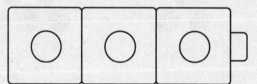

_____

- - - - - - - - -

_____

_____

- - - - - - - - -

_____

DIRECTIONS **1.** How many cubes are there? Write the number. Model a cube train that has a number of cubes greater than 4. Draw the cube train. Write how many. Compare the cube trains by matching. Tell a friend about the cube trains. **2.** How many cubes are there? Write the number. Model a cube train that has a number of cubes less than 3. Draw the cube train. Write how many. Compare the cube trains by matching. Tell a friend about the cube trains.

Lesson Check

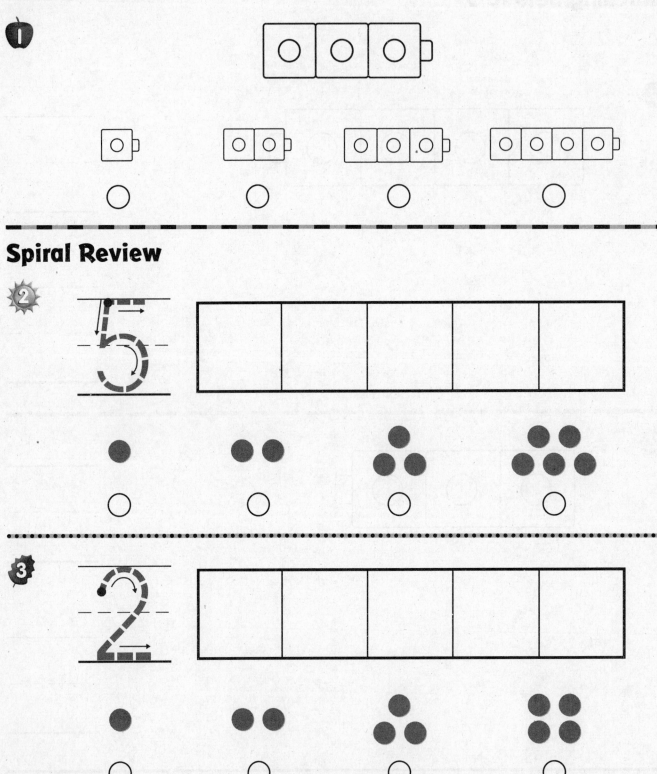

Spiral Review

DIRECTIONS 1. Which cube train has a number of cubes greater than 3?
Mark under your answer. (Lesson 2.4) 2–3. Trace the number. How many counters would
you place in the five frame to show the number? Mark under your answer. (Lessons 1.5, 1.1)

P34 thirty-four

Compare by Counting Sets to 5

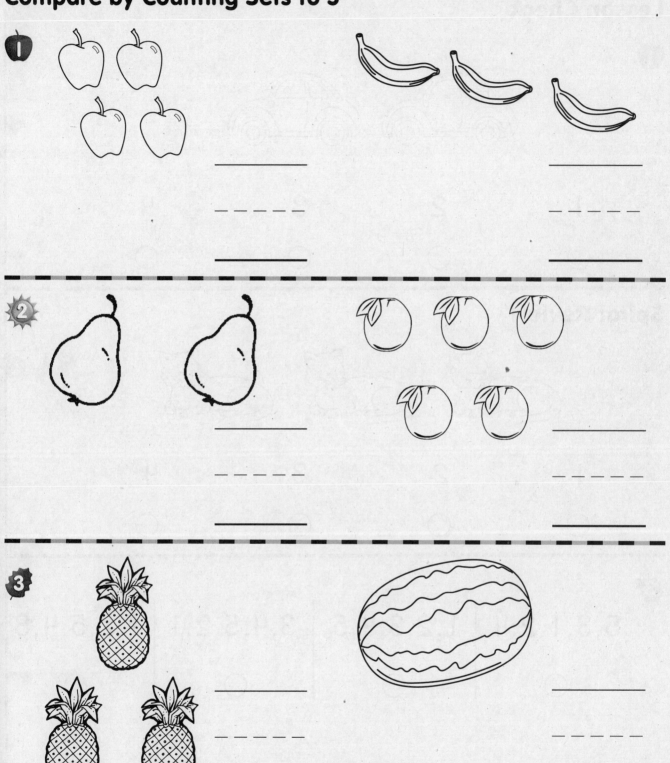

DIRECTIONS 1–2. Count how many objects are in each set. Write the numbers. Compare the numbers. Circle the number that is greater. 3. Count how many objects are in each set. Write the numbers. Compare the numbers. Circle the number that is less.

Lesson Check

1	2	3	4
○	○	○	○

Spiral Review

1	2	3	4
○	○	○	○

5, 3, 1, 2, 4	1, 2, 3, 4, 5	3, 4, 5, 2, 1	1, 2, 5, 4, 3
○	○	○	○

DIRECTIONS **1.** Mark under the number that is less than the number of cars. **(Lesson 2.5)** **2.** Count and tell how many cats. Mark under your answer. **(Lesson 1.4)** **3.** Which set of numbers is in order? Mark under your answer. **(Lesson 1.8)**

Chapter 2 Extra Practice

Lessons 2.1 – 2.2 (pp. 61–68) ·

_____   _____

_ _ _ _ _ _ _   _ _ _ _ _ _ _

_____   _____

● ●

   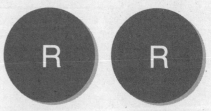

   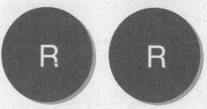

_____   _____

_ _ _ _ _ _ _   _ _ _ _ _ _ _

_____   _____

DIRECTIONS   1. Compare the sets of objects. Is the number of lunch boxes greater than, less than, or the same as the number of backpacks? Count how many lunch boxes. Write the number. Count how many backpacks. Write the number. Tell a friend what you know about the number of objects in each set.   **2.** Place counters as shown. Y is for yellow, and R is for red. Count and tell how many in each set. Write the numbers. Circle the number that is greater.

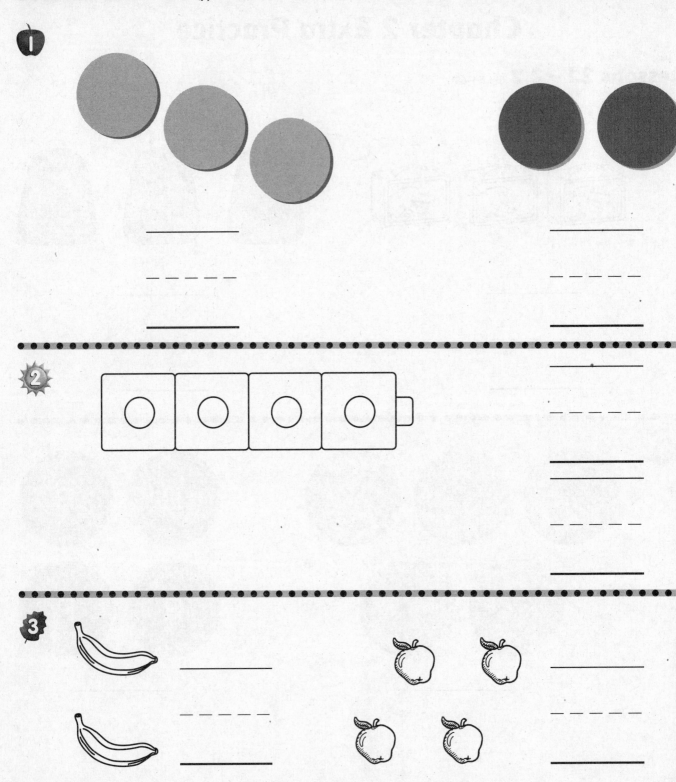

DIRECTIONS **I.** Count and tell how many in each set. Write the numbers. Compare the numbers. Circle the number that is less. **2.** How many cubes are there? Write the number. Model a cube train that has a number of cubes less than 4. Draw the cube train. Write how many. Compare the cube trains by matching. Tell a friend about the cube trains. **3.** Count how many objects in each set. Write the numbers. Compare the numbers. Circle the number that is greater.

School-Home Letter

Dear Family,

My class started Chapter 3 this week. In this chapter, I will learn how to show, count, and write numbers 6 to 9.

Love, _____

Vocabulary

six one more than five

eight one more than seven

Home Activity

Pour salt or sand into a cookie sheet or baking dish. Pick a number from 6 to 9 and have your child draw the number in the salt or sand. Then ask your child to draw circles to match that number. Shake to erase and begin again!

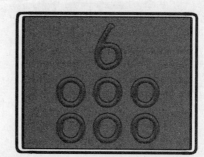

Literature

Look for this book in the library. You and your child will enjoy this fun story that provides reinforcement of counting concepts.

Seven Scary Monsters by Mary Beth Lundgren. Clarion Books, 2003.

Carta para la casa

Querida familia:

Mi clase comenzó el Capítulo 3 esta semana. En este capítulo, aprenderé cómo mostrar, contar y escribir números del 6 al 9.

Con cariño, _____

Vocabulario

seis uno más que cinco

ocho uno más que siete

Actividad para la casa

Ponga sal o arena en una fuente para horno. Elija un número del 6 al 9 y pídale a su hijo que dibuje el número en la sal o la arena. Luego, pídale que dibuje el mismo número de círculos. Mezcle para borrar y ¡comiencen de nuevo!

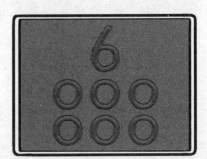

Literatura

Busque este libro en la biblioteca. Usted y su hijo disfrutarán de este cuento divertido que proporciona un refuerzo para los conceptos de contar.

Seven Scary Monsters por Mary Beth Lundgren. Clarion Books, 2003.

Name _____

Model and Count 6

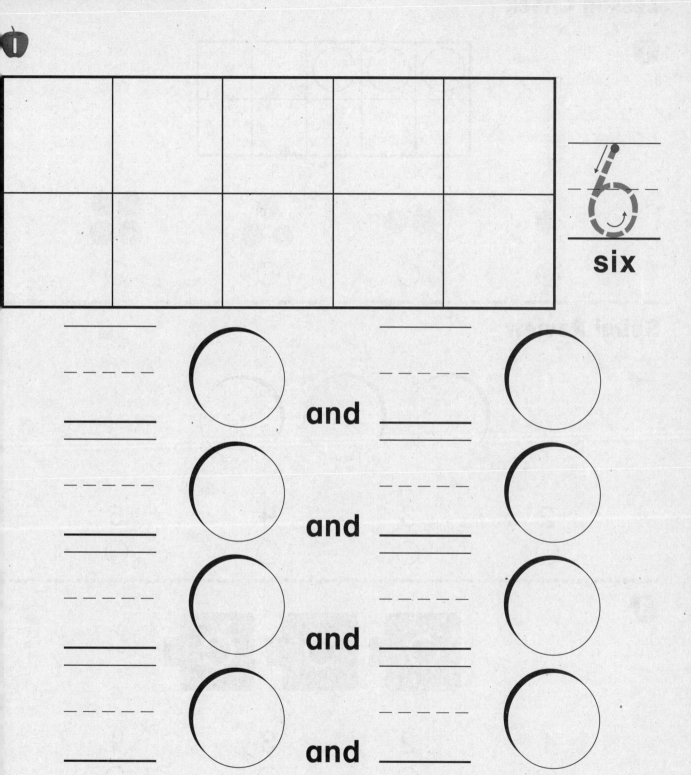

six

and

and

and

and

DIRECTIONS 1. Trace the number 6. Use two-color counters to model the different ways to make 6. Color to show the counters below. Write to show some pairs of numbers that make 6.

Lesson Check

1

Spiral Review

2

2   3   4   5

○   ○   ○   ○

3

1   2   3   4

○   ○   ○   ○

DIRECTIONS **1.** How many more counters would you place to model a way to make 6? Mark under your answer. **(Lesson 3.1)** **2.** Mark under the number that is less than the number of counters. **(Lesson 2.3)** **3.** How many cubes are there? Mark under your answer. **(Lesson 1.4)**

Name _____

Read and Write to 6

 1

6
six

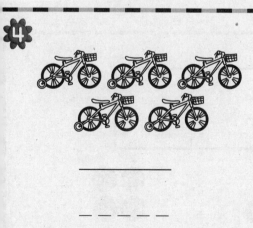 **2**

 **3**

_____

- - - - - - -

_____

4

_____

- - - - - - -

_____

5

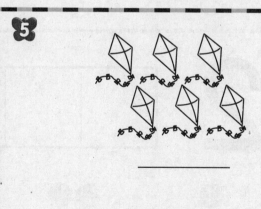

_____

- - - - - - -

_____

DIRECTIONS I. Say the number. Trace the numbers.
2–5. Count and tell how many. Write the number.

Lesson Check

3	4	5	6
○	○	○	○

Spiral Review

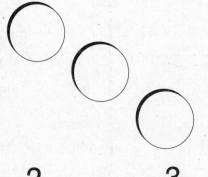

1	2	3	4
○	○	○	○

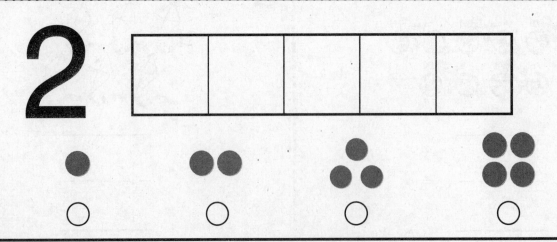

DIRECTIONS **1.** How many school buses are there? Mark under your answer.
(Lesson 3.2) **2.** Mark under the number that is greater than the number of
counters. **(Lesson 2.2)** **3.** How many counters would you place in the five frame to
show the number? Mark under your answer. **(Lesson 1.1)**

Name _____

Model and Count 7

①

7
seven

___ ___ ◯   **and**   ___ ___ ◯

___ ___ ◯   **and**   ___ ___ ◯

___ ___ ◯   **and**   ___ ___ ◯

___ ___ ◯   **and**   ___ ___ ◯

DIRECTIONS   1. Trace the number 7. Use two-color counters to model the different ways to make 7. Color to show the counters below. Write to show some pairs of numbers that make 7.

Lesson Check

1

Spiral Review

2

1	2	3	4
○	○	○	○

3

1	2	3	4
○	○	○	○

DIRECTIONS   **I.** How many more counters would you place to model a way to make 7? Mark under your answer. **(Lesson 3.3)**   **2.** Mark under the number that is less than the number of counters. **(Lesson 2.3)**   **3.** How many birds are there? Mark under your answer. **(Lesson 1.4)**

Read and Write to 7

7
seven

7 7 7 7 7 7 7

DIRECTIONS 1. Say the number. Trace the numbers.
2–5. Count and tell how many. Write the number.

Lesson Check

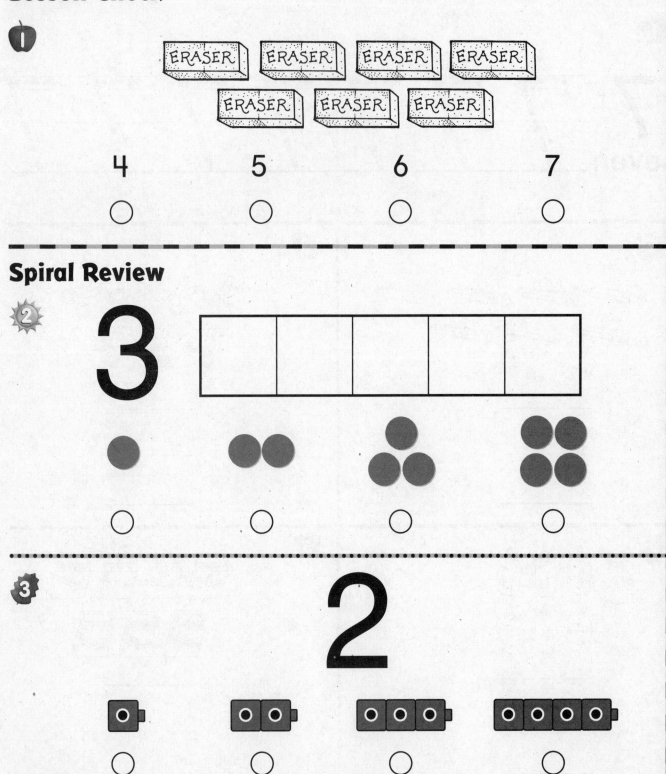

1

ERASER ERASER ERASER ERASER

ERASER ERASER ERASER

4   5   6   7

○   ○   ○   ○

Spiral Review

2

3

○   ○   ○   ○

3

2

○   ○   ○   ○

DIRECTIONS 1. Count and tell how many erasers. Mark under your answer.
(Lesson 3.4)   2. How many counters would you place in the five frame to show the
number? Mark under your answer. (Lesson 1.3)   3. Which set shows the number?
Mark under your answer. (Lesson 1.2)

Name _____

Model and Count 8

8
eight

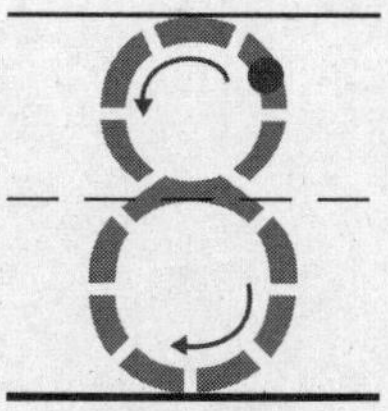

and

and

and

and

DIRECTIONS 1. Trace the number 8. Use two-color counters to model the different ways to make 8. Color to show the counters below. Write to show some pairs of numbers that make 8.

Lesson Check

1

2

Spiral Review

3

DIRECTIONS **1.** How many more counters would you place to model a way to make 8? Mark under your answer. **(Lesson 3.5)** **2.** Which cube train has a number of cubes greater than 4? Mark under your answer. **(Lesson 2.4)** **3.** Count and tell how many stop signs. Mark under your answer. **(Lesson 1.2)**

Read and Write to 8

8

eight

8 8 8 8 8 8 8

_____

- - - - - - -

_____

_____

- - - - - - -

_____

_____

- - - - - - -

_____

_____

- - - - - - -

_____

DIRECTIONS 1. Say the number. Trace the numbers.
2–5. Count and tell how many. Write the number.

8	7	6	4
○	○	○	○

Spiral Review

2	3	4	5
○	○	○	○

two	three	four	five
○	○	○	○

DIRECTIONS **1.** Count and tell how many bees. Mark under your answer. **(Lesson 3.6)**
2. Mark under the number that is greater than the number of counters. **(Lesson 2.2)**
3. Count and tell how many beetles. Mark under your answer. **(Lesson 1.6)**

Model and Count 9

nine

and

and

and

and

and

and

and

and

DIRECTIONS 1. Trace the number 9. Use two-color counters to model the different ways to make 9. Color to show the counters below. Write to show some pairs of numbers that make 9.

Lesson Check

1

Spiral Review

2

1   2   3   4

○   ○   ○   ○

3

1   2   3   4

○   ○   ○   ○

DIRECTIONS 1. How many more counters would you place to model a way to make 9? Mark under your answer. (Lesson 3.7) 2. Mark under the number that is greater than the number of cats. (Lesson 2.5) 3. How many counters are there? Mark under your answer. (Lesson 1.4)

Read and Write to 9

 1

9
nine

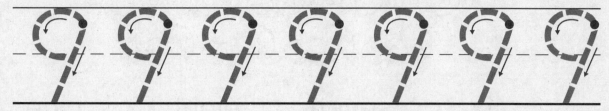

 2

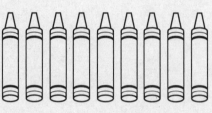

_____

- - - - - - - -

_____

3

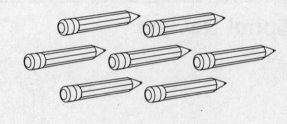

_____

- - - - - - - -

_____

 4

_____

- - - - - - - -

_____

5

_____

- - - - - - - -

_____

DIRECTIONS I. Say the number. Trace the numbers.
2–5. Count and tell how many. Write the number.

Lesson Check

six	seven	eight	nine
◯	◯	◯	◯

Spiral Review

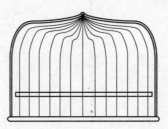

0	1	2	3
◯	◯	◯	◯

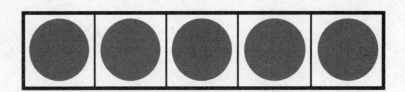

2	3	4	5
◯	◯	◯	◯

DIRECTIONS 1. Count and tell how many squirrels. Mark under your answer.
(Lesson 3.8) 2. How many birds are in the cage? Mark under your answer.
(Lesson 1.10) 3. How many counters are there? Mark under your answer. (Lesson 1.6)

Name _____

Problem Solving • Numbers to 9

①

_____
- - - - - - - -
_____
_____
- - - - - - - -
_____

②

_____
- - - - - - - -
_____
_____
- - - - - - - -
_____

DIRECTIONS 1. Sally has six flowers. Three of the flowers are yellow. The rest are red. How many are yellow? Draw the flowers. Write the number beside each set of flowers. **2.** Tim has seven acorns. Don has a number of acorns that is two less than 7. How many acorns does Don have? Draw the acorns. Write the numbers.

Chapter 3

Lesson Check

2	3	5	7
◯	◯	◯	◯

Spiral Review

2	3	4	5
◯	◯	◯	◯

2	3	4	5
◯	◯	◯	◯

DIRECTIONS 1. The house has five doors. The number of windows is two more than 5. How many windows are there? Mark under your answer. **(Lesson 3.9) 2.** Count and tell how many books. Mark your answer. **(Lesson 1.6) 3.** Mark under the number that is greater than the number of turtles. **(Lesson 2.5)**

Name _____

Chapter 3 Extra Practice

Lessons 3.1 – 3.6 (pp. 89–112) . . . .

1

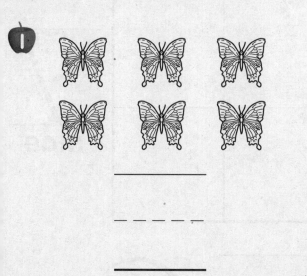

_____

- - - - - - -

_____

2

_____

- - - - - - -

_____

3

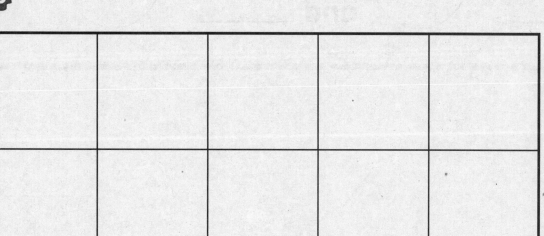

8

eight

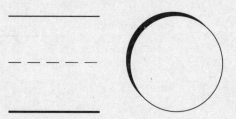 **and**

DIRECTIONS   1–2. Count and tell how many. Write the number.   **3.** Trace the number 8. Use two-color counters to model a way to make 8. Color to show the counters below. Write to show a pair of numbers that makes 8.

①

9

nine

_____

_____ ⬤ **and** _____ ⬤

②

_____

_____

_____

_____

_____

_____

_____

DIRECTIONS **1.** Trace the number 9. Use two-color counters to model a way to make 9. Color to show the counters below. Write to show a pair of numbers that makes 9. **2.** Roy has seven spoons. Ken has a number of spoons two greater than 7. Draw the spoons. Write the numbers.

School-Home
Letter

Dear Family,

My class started Chapter 4 this week. In this chapter, I will learn how to show and compare numbers to 10.

Love, _____

Vocabulary

ten one more than nine

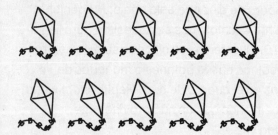

Home Activity

Place one button or penny in the ten frame below. Ask your child how many more are needed to make 10. Count aloud with your child as he or she places nine more buttons or pennies in the ten frame. Repeat the activity, starting with a different number each time.

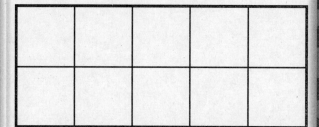

Literature

Look for these books in the library. You and your child will enjoy these fun stories while learning more about the numbers 6 to 10.

Feast for 10
by Cathryn Falwell.
Clarion Books, 1993.

Ten Black Dots
by Donald Crews.
Greenwillow Books, 1995.

Carta
para la casa

Querida familia:

Mi clase comenzó el Capítulo 4 esta semana. En este capítulo, aprenderé mostrar y comparar números hasta el 10.

Con cariño, _____

Vocabulario

diez uno más que nueve

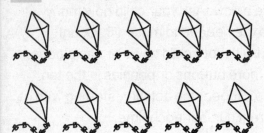

Actividad para la casa

Ponga un botón o una moneda de 1¢ en el cuadro de diez que está abajo. Pregúntele a su hijo cuántos más se necesitan para llegar a 10. Cuente en voz alta con su hijo mientras él coloca nueve botones o monedas de 1¢ más en el cuadro de diez. Repita la actividad y comience con un número diferente cada vez.

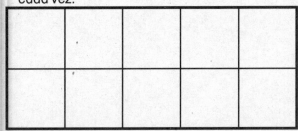

Literatura

Busquen estos libros en la biblioteca. Usted y su hijo se divertirán leyendo estos cuentos mientras aprenden más sobre los números del 6 al 10.

Feast for 10 by Cathryn Falwell. Clarion Books, 1993.

Ten Black Dots by Donald Crews. Greenwillow Books, 1995.

Name _____

Model and Count 10

10
ten

_____
- - - - -
_____ ◯ **and** - - - - - ◯ _____

_____
- - - - - ◯ **and** - - - - - ◯
_____   _____

- - - - - ◯ **and** - - - - - ◯
_____   _____

_____
- - - - - ◯ **and** - - - - - ◯
_____   _____

DIRECTIONS Trace the number. Use counters to model the different ways to make 10.
Color to show the counters below. Write to show some pairs of numbers that make 10.

Lesson Check

1.

○   ○   ○   ○

Spiral Review

2.

○   ○   ○   ○

3.

4   3   2   1

○   ○   ○   ○

DIRECTIONS **1.** How many more counters would you place to model a way to make 10? Mark under your answer. (Lesson 4.1) **2.** Mark under the set that has the same number of objects as the set of kites. (Lesson 2.1) **3.** Count and tell how many coats. Mark under your answer. (Lesson 1.2)

Read and Write to 10

1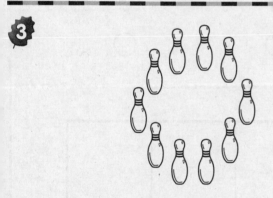

10
ten

2

_____

- - - - - - - - -

_____

3

_____

- - - - - - - - -

_____

4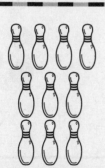

_____

- - - - - - - - -

_____

DIRECTIONS **1.** Say the number. Trace the numbers.
2–4. Count and tell how many. Write the number.

Lesson Check

seven	eight	nine	ten
○	○	○	○

Spiral Review

2	3	4	5
○	○	○	○

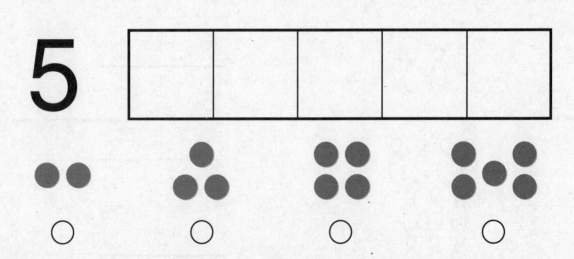

DIRECTIONS 1. Count and tell how many ears of corn. Mark under your answer.
(Lesson 4.2) 2. Mark under the number that is less than the number of counters.
(Lesson 2.3) 3. How many counters would you place in the five frame to show the number?
Mark under your answer. (Lesson 1.5)

Algebra • Ways to Make 10

1.

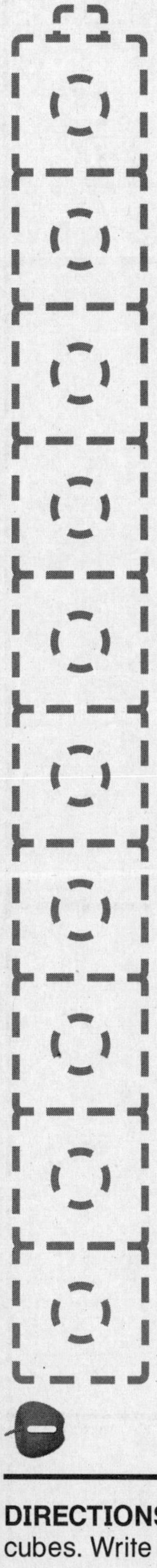

cubes

10

red

blue

7

2.

cubes

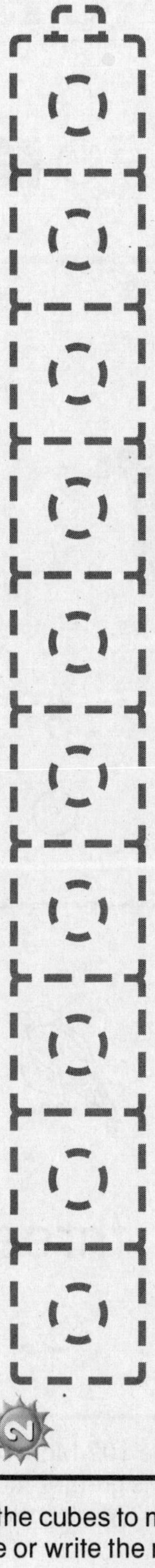

red

blue

6

3.

cubes

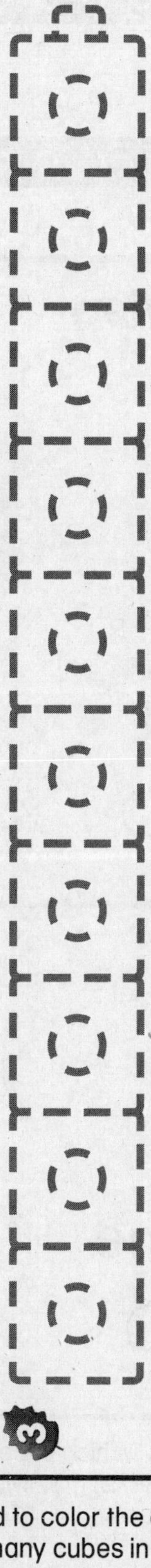

red

blue

2

DIRECTIONS 1–3. Use blue to color the cubes to match the number. Use red to color the other cubes. Write how many red cubes. Trace or write the number that shows how many cubes in all.

Lesson Check

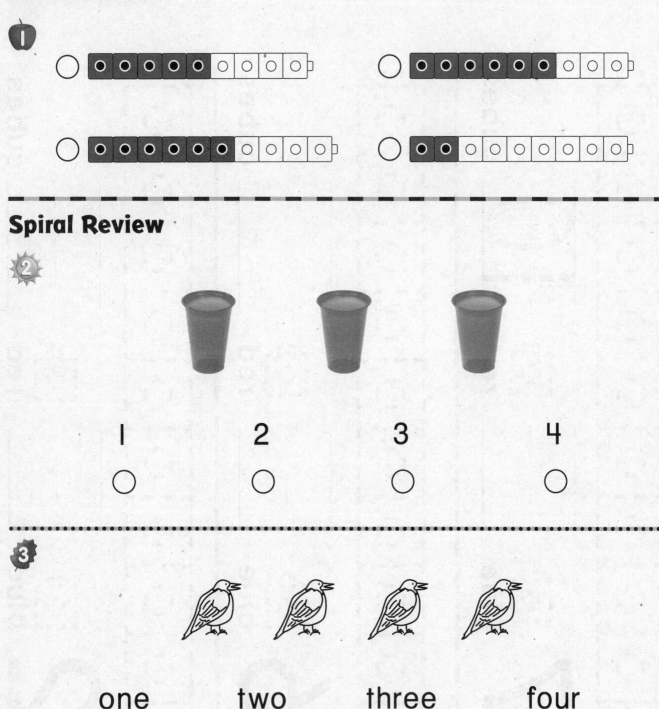

Spiral Review

1   2   3   4

○   ○   ○   ○

one   two   three   four

○   ○   ○   ○

DIRECTIONS 1. Which cube train shows a way to make 10? Mark beside your answer. **(Lesson 4.3)** 2. Mark under the number that is greater than the number of cups. **(Lesson 2.5)** 3. How many birds are there? Mark under your answer. **(Lesson 1.4)**

Count and Order to 10

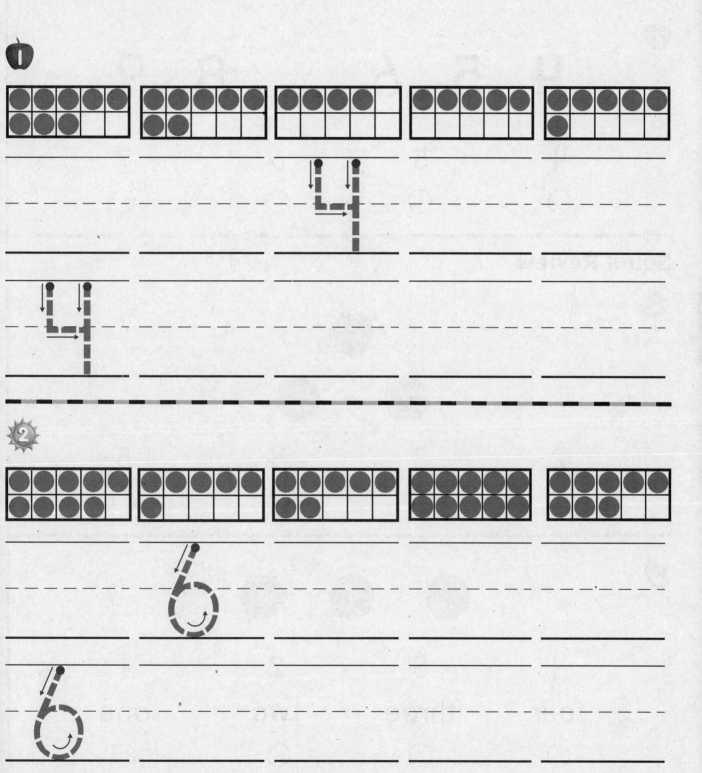

DIRECTIONS 1–2. Count the dots in the ten frames. Trace or write the numbers. Write the numbers in order as you count forward from the dashed number.

Lesson Check

❶

$$4 \quad 5 \quad 6 \quad \underline{} \quad 8 \quad 9$$

4	5	6	7
○	○	○	○

Spiral Review

②

2	3	4	5
○	○	○	○

③

4	3	2	I
four	three	two	one
○	○	○	○

DIRECTIONS I. Count forward. Mark under the number that fills the space. (Lesson 4.4) 2. Mark under the number that is less than the number of counters. (Lesson 2.3) 3. How many counters are there? Mark under your answer. (Lesson 1.4)

Name _____

Problem Solving • Compare by Matching Sets to 10

1

_____
- - - - - -
_____

_____
- - - - - -
_____

2

_____
- - - - - -
_____

_____
- - - - - -
_____

DIRECTIONS **1.** Kim has 7 red balloons. Jake has 3 blue balloons. Who has fewer balloons? Use cube trains to model the sets of balloons. Compare the cube trains. Write how many. Circle the number that is less. **2.** Meg has 8 red beads. Beni has 5 blue beads. Who has more beads? Use cube trains to model the sets of beads. Compare the cube trains by matching. Draw and color the cube trains. Write how many. Circle the number that is greater.

Chapter 4   seventy-one **P71**

Lesson Check

1

○ ▢▢▢▢▢▢▢

○ ▢▢▢▢▢▢▢

○ ▢▢▢▢▢▢▢

○ ▢▢▢▢▢▢

Spiral Review

2

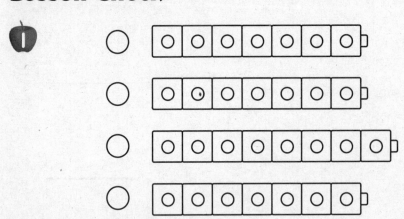

1	2	3	4
○	○	○	○

3

5

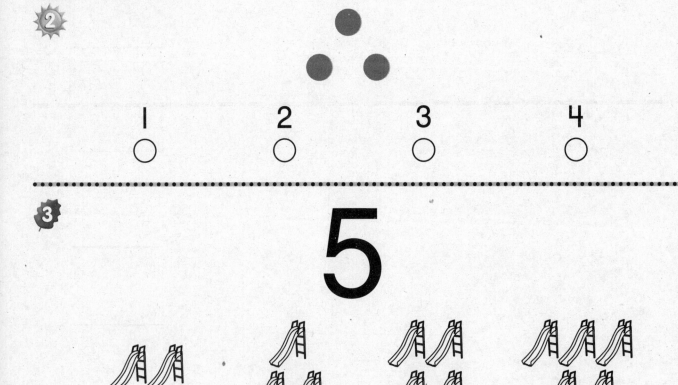

○　　　○　　　○　　　○

DIRECTIONS 1. Compare the cube trains by matching. Mark beside the cube train that has a greater number of cubes. **(Lesson 4.5)** 2. Mark under the number that is greater than the number of counters. **(Lesson 2.2)** 3. Which set shows the number? Mark under your answer. **(Lesson 1.6)**

Compare by Counting Sets to 10

1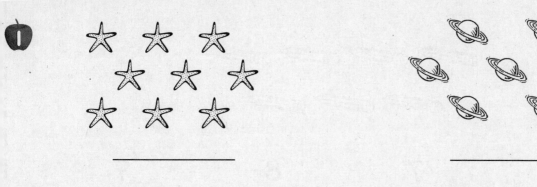

_____

- - - - - - -

_____

2

_____   _____

- - - - - - -   - - - - - - -

_____   _____

3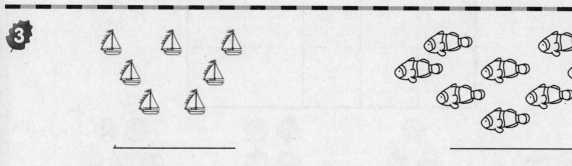

_____   _____

- - - - - - -   - - - - - - -

_____   _____

DIRECTIONS Count how many in each set. Write the number of objects in each set. Compare the numbers. **1–2.** Circle the number that is less. **3.** Circle the number that is greater.

Lesson Check

1

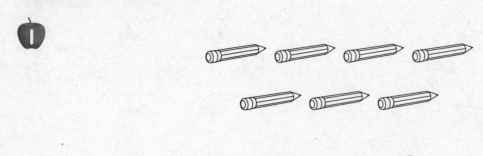

6 ○   7 ○   8 ○   9 ○

Spiral Review

2

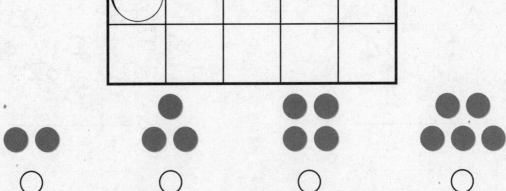

5 ○   6 ○   7 ○   8 ○

3

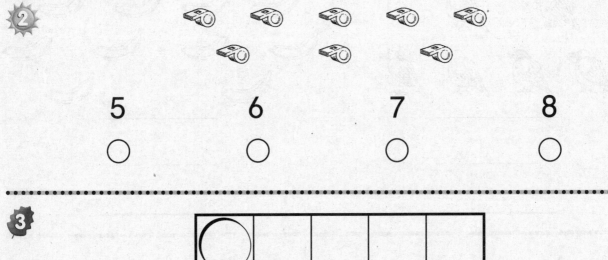

○   ○   ○   ○

DIRECTIONS **1.** Mark under the number that is less than the number of pencils.
(Lesson 4.6)   **2.** Count and tell how many whistles. Mark under your answer. (Lesson 3.6)
3. How many more counters would you place to model a way to make 6? (Lesson 3.1)

Compare Two Numbers

1.   8   5

---

2.   10   7

---

3.   6   9

---

4.   4   6

---

5.   8   7

---

6.   5   3

DIRECTIONS   **1–3.** Look at the numbers. Think about the counting order as you compare the numbers. Circle the greater number.   **4–6.** Look at the numbers. Think about the counting order as you compare the numbers. Circle the number that is less.

Lesson Check

1

$$7$$

6   8   I   5

○   ○   ○   ○

Spiral Review

2

six   seven   eight   nine

○   ○   ○   ○

DIRECTIONS 1. Which number is greater than 7? Mark under your answer. (Lesson 4.7)
2. How many more counters would you place to model a way to make 8? Mark under your answer. (Lesson 3.5) 3. Count and tell how many birds. Mark under your answer. (Lesson 3.8)

Chapter 4 Extra Practice

Lessons 4.1–4.4 (pp. 133–147) • • • • • • • • • • • • • • • •

 1

_____

- - - - - - - - - - -

2

3   **blue** _____ **red**   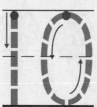 **cubes**

3

6, 8, 5, 7, 9

5 ___ ___ ___ ___

- - - - - - - - - - -

DIRECTIONS   **1.** Count and tell how many balloons. Write the number.
2. Use blue to color the cubes to match the number. Use red to color the other cubes. Write how many red cubes. Trace the number that shows how many cubes in all.   **3.** Write the numbers in order as you count forward from 5.

1

2

8   4

DIRECTIONS **1.** Pam has 9 red crayons. Alex has 7 blue crayons. Who has more crayons? Use cube trains to model the sets of crayons. Compare the cube trains by matching. Draw and color the cube trains. Write how many. Circle the number that is greater. **2.** Count how many in each set. Write the number of objects in each set. Compare the numbers. Circle the number that is less. **3.** Think about the counting order as you compare the numbers. Circle the greater number.

School-Home Letter

Dear Family,

My class started Chapter 5 this week. In this chapter, I will learn how to show addition.

Love, _____

Vocabulary

add to put together one set with another set

$$3 + 2 = 5$$

plus (+) a symbol that shows addition

plus
↓
$$3 + 2 = 5$$

Home Activity

Invite your child to act out addition word problems. For example, your child can show you four socks, add two more socks, and then tell you the addition sentence.

$$4 + 2 = 6$$

Literature

Look for these books at the library. You and your child will enjoy counting and adding objects in these interactive books.

Rooster's Off to See the World by Eric Carle. Simon & Schuster, 1991.

Anno's Counting Book by Mitsumasa Anno. HarperCollins, 1986.

Carta
para la casa

Querida familia:

Mi clase comenzó el Capítulo 5 esta semana. En este capítulo aprenderé todo sobre la suma.

Con cariño, _____

Vocabulario

sumar agregar un conjunto a otro

más (+) signo que indica suma

más
↓
$3 + 2 = 5$

Actividad para la casa

Anime a su hijo a representar problemas de suma. Por ejemplo, puede mostrar cuatro calcetines, agregar dos calcetines más y luego decir el enunciado de la suma.

$4 + 2 = 6$

Busquen otros objetos que puedan usarse para representar cuentos de resta.

Literatura

Busquen estos libros en la biblioteca. Usted y su hijo disfrutarán estos libros interactivos que sirven para reforzar las destrezas de suma.

Rooster's Off to See the World
por Eric Carle. Simon & Schuster, 1991.

Anno's Counting Book
by Mitsumasa Anno. HarperCollins, 1986.

Addition: Add To

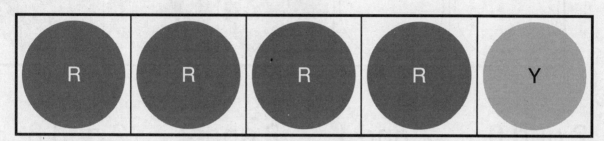

_____   _____

- - - - -   **and**   - - - - -

_____   _____

_____

- - - - -

_____

DIRECTIONS   **1.** There are four red counters in the five frame. One yellow counter is added. R is for red, and Y is for yellow. How many of each color counter? Write the numbers.   **2.** Write the number that shows how many counters are in the five frame now.

Chapter 5

eighty-one **P81**

Lesson Check

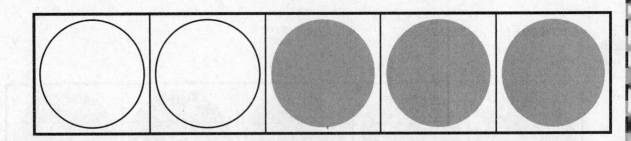

2 and 0   2 and 1   2 and 2   2 and 3

○   ○   ○   ○

Spiral Review

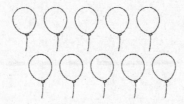

7   8   9   10

○   ○   ○   ○

2   3   4   5

○   ○   ○   ○

DIRECTIONS   **1.** Which shows the gray counters being added to the five frame? Mark under your answer. **(Lesson 5.1)**   **2.** Count and tell how many balloons. Mark under your answer. **(Lesson 4.2)**   **3.** Mark under the number that is less than the number of shells. **(Lesson 2.5)**

Addition: Put Together

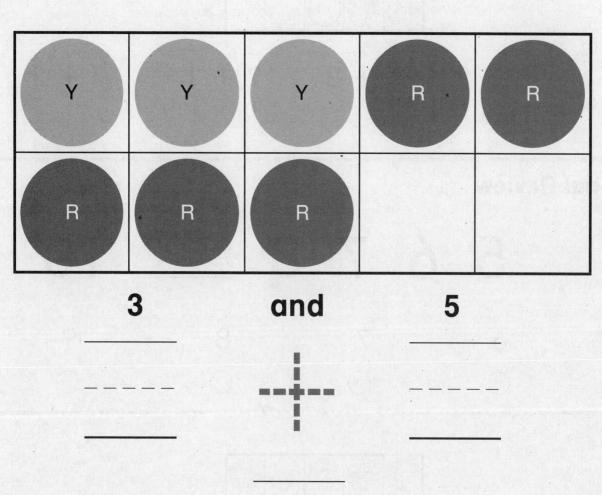

3   and   5

DIRECTIONS   Roy has three yellow counters and five red counters. How
many counters does he have in all? **1.** Place counters in the ten frame to
model the sets that are put together. Y is for yellow, and R is for red. Write
the numbers and trace the symbol. Write the number to show how many in all.

Lesson Check

1

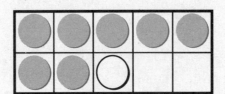

$5 + 2$ ○ $5 + 3$ ○ $7 + 1$ ○ $7 + 2$ ○

Spiral Review

5   6   7   8   ___   10

6 ○   7 ○   8 ○   9 ○

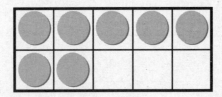

5 ○   6 ○   7 ○   9 ○

DIRECTIONS 1. Which numbers show the sets that are put together? Mark under your answer. **(Lesson 5.2)** 2. Count forward. Mark under the number that fills the space. **(Lesson 4.4)** 3. Meg has seven counters. Paul has a number of counters two less than seven. Mark under the number that shows how many counters Paul has. **(Lesson 3.9)**

Problem Solving • Act Out
Addition Problems

1

$$4 + 1 = \underline{\qquad}$$

2

$$3 + 2 = \underline{\qquad}$$

DIRECTIONS   1–2. Tell an addition word problem about the children. Trace the numbers and the symbols. Write the number that shows how many children in all.

Lesson Check

$$3 + 2 = \underline{\qquad}$$

5	4	3	2
○	○	○	○

Spiral Review

three	four	five	six
○	○	○	○

○ ○ ○ ○

DIRECTIONS **1.** How many cats are there in all? Mark under your answer. **(Lesson 5.3)**
2. Count and tell how many tigers. Mark under your answer. **(Lesson 3.2)** **3.** Mark under
the set that has the same number of objects. **(Lesson 2.1)**

Name _____

Algebra • Model and Draw
Addition Problems

1

- - - - - -

2

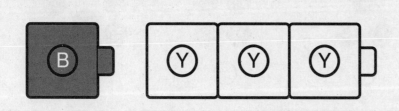

DIRECTIONS 1–2. Place cubes as shown. B is for blue, and Y is for yellow. Tell an addition word problem. Model to show the cubes put together. Draw the cube train. Trace and write to complete the addition sentence.

○ 2 + 1 = 3   ○ 3 + 1 = 4

○ 2 + 3 = 5   ○ 3 + 2 = 5

Spiral Review

  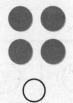 

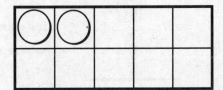

   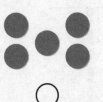

DIRECTIONS **1.** Which addition sentence shows the cubes being put together? Mark beside your answer. **(Lesson 5.4)** **2.** How many more counters would you place to model a way to make 7? Mark under your answer. **(Lesson 3.3)** **3.** Mark under the set that shows the number. **(Lesson 1.4)**

Algebra • Write Addition Sentences for 10

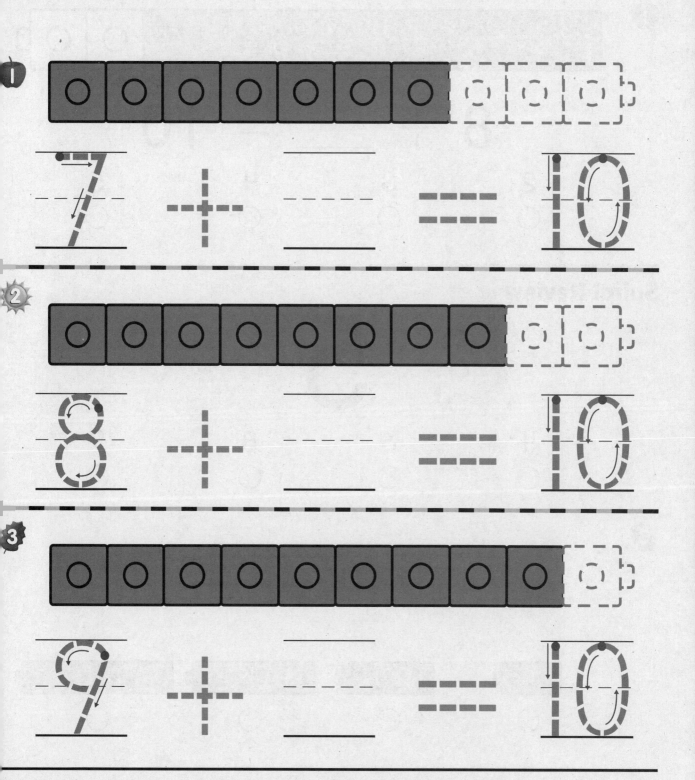

1 7 + ___ = 10

2 8 + ___ = 10

3 9 + ___ = 10

DIRECTIONS 1–3. Look at the cube train. How many gray cubes do you see? How many blue cubes do you need to add to make 10? Use blue to color those cubes. Write and trace to show this as an addition sentence.

Lesson Check

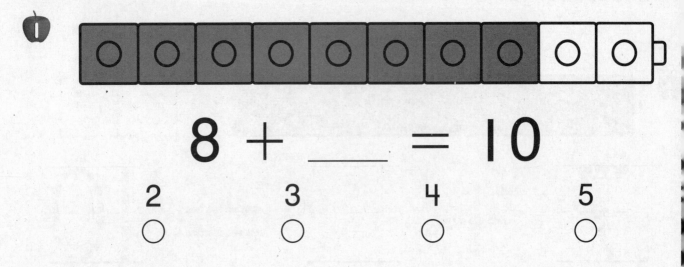

$$8 + \underline{\quad} = 10$$

2   3   4   5
○   ○   ○   ○

Spiral Review

5

4   6   8   9
○   ○   ○   ○

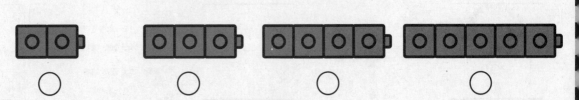

○   ○   ○   ○

DIRECTIONS   1. Mark under the number that makes 10 when put together with the given number. **(Lesson 5.5)   2.** Which number is less than 5? Mark under your answer. **(Lesson 4.7)   3.** Which cube train has the same number of cubes? Mark under your answer. **(Lesson 2.4)**

Algebra • Write Addition Sentences

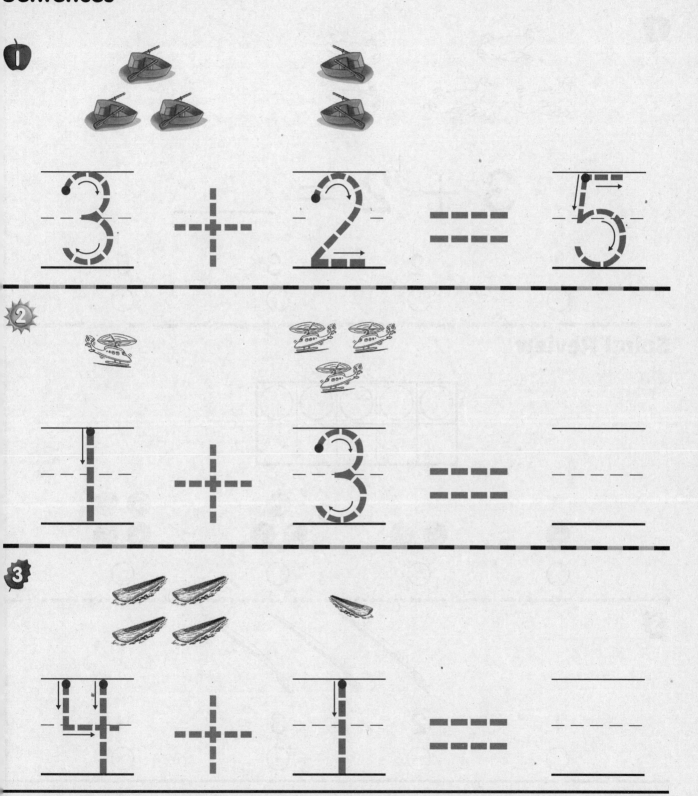

DIRECTIONS 1–3. Tell an addition word problem about the sets. How many are in the set you start with? How many are being added to the set? How many are there now? Write and trace to complete the addition sentence.

Lesson Check

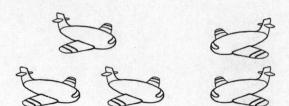

$$3 + 2 = \underline{\quad}$$

1   2   3   5
○   ○   ○   ○

Spiral Review

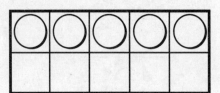

○   ○   ○   ○

1   2   3   4
○   ○   ○   ○

DIRECTIONS **1.** Which number completes the addition sentence about the sets of airplanes? Mark under your answer. **(Lesson 5.6)** **2.** How many more counters would you place to model a way to make 8? Mark under your answer. **(Lesson 3.5)** **3.** How many paintbrushes are there? Mark under your answer. **(Lesson 1.4)**

Algebra • Write More
Addition Sentences

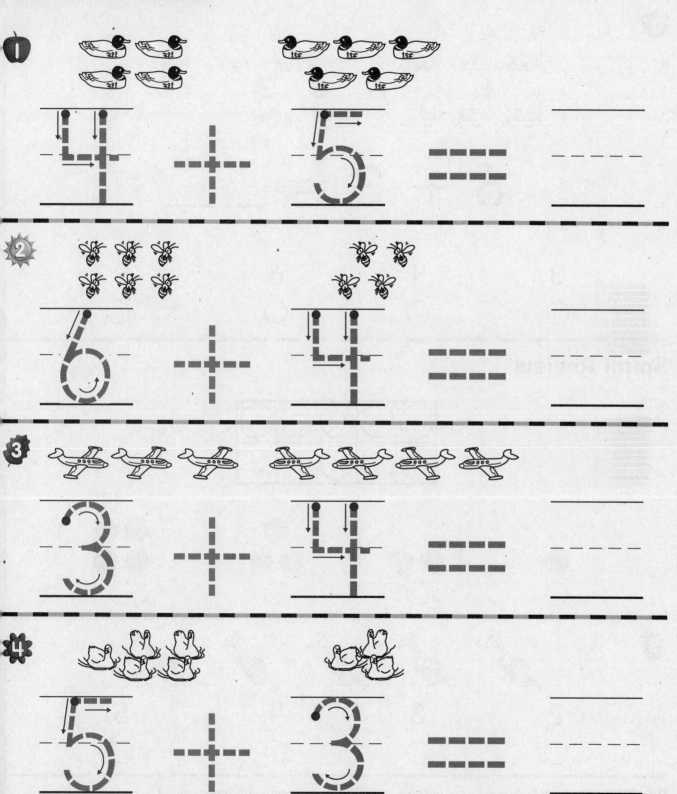

1   4 + 5 == _____

2   6 + 4 == _____

3   3 + 4 == _____

4   5 + 3 == _____

DIRECTIONS 1–4. Tell an addition word problem about the sets. How many are in the set to start with? How many are joining? How many are there now? Write and trace to complete the addition sentence.

Chapter 5   ninety-three **P93**

Lesson Check

$$6 + 3 = \underline{\qquad}$$

3	4	6	9
○	○	○	○

Spiral Review

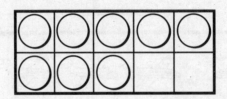

○	○	○	○

2	3	4	5
○	○	○	○

DIRECTIONS 1. Which number completes the addition sentence about the sets of dogs? Mark under your answer. (Lesson 5.7) 2. How many more counters would you place to model a way to make 9? Mark under your answer. (Lesson 3.7) 3. Count and tell how many trumpets. Mark under your answer. (Lesson 1.6)

Algebra • Number Pairs to 5

1

3   === ____ ___ + ___

2

4   === ____ ___ + ___

3

5   === ____ ___ + ___

DIRECTIONS 1–3. Look at the number at the beginning of the addition
sentence. Place two colors of cubes on the cube train to show a number
pair for that number. Complete the addition sentence to show a number
pair. Color the cube train to match the addition sentence.

Lesson Check

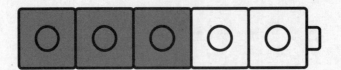

○   5 = 1 + 4   ○   6 = 1 + 5

○   5 = 3 + 2   ○   6 = 2 + 4

Spiral Review

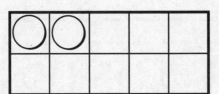

5   6   7   8

○   ○   ○   ○

○   ○   ○   ○

DIRECTIONS   **1.** Which addition sentence shows a pair of numbers that matches the cube train? Mark beside your answer. **(Lesson 5.8)**   **2.** Mark under the number that is greater than the number of turtles. **(Lesson 4.6)**   **3.** How many more counters would you place to model a way to make 6? Mark under your answer **(Lesson 3.1)**

P96   ninety-six

Algebra • Number Pairs for 6 and 7

1

6 === _ _ _ _ + _ _ _ _

2

7 === _ _ _ _ + _ _ _ _

DIRECTIONS 1–2. Look at the number at the beginning of the addition sentence. Place two colors of cubes on the cube train to show a number pair for that number. Complete the addition sentence to show a number pair. Color the cube train to match the addition sentence.

Lesson Check

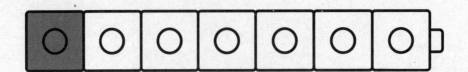

○ 6 = 1 + 5   ○ 7 = 1 + 6

○ 6 = 2 + 4   ○ 7 = 3 + 4

Spiral Review

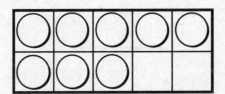

○   ○   ○   ○

four   five   six   seven

○   ○   ○   ○

DIRECTIONS   **1.** Which addition sentence shows a pair of numbers that matches the cube train? Mark beside your answer. **(Lesson 5.9)**   **2.** How many more counters would you place to model a way to make 10? Mark under your answer. **(Lesson 4.1)**   **3.** Count and tell how many hats. Mark under your answer. **(Lesson 3.4)**

Algebra • Number Pairs for 8

① □ □ □ □ □ □ □ □

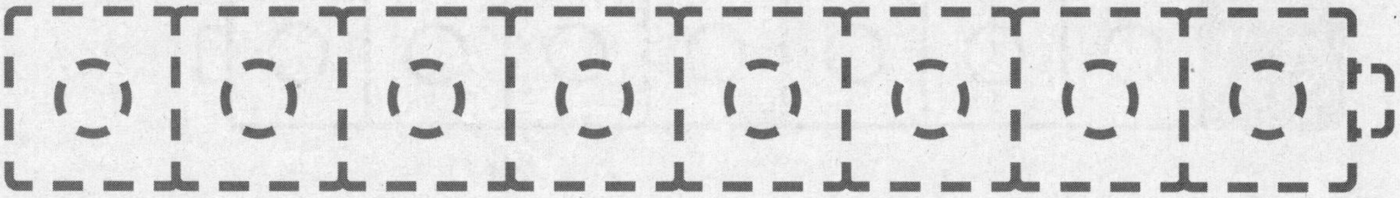

8 ═ ____ ✛ ____

② **8** ═ ____ ✛ ____

③ **8** ═ ____ ✛ ____

④ **8** ═ ____ ✛ ____

DIRECTIONS Use two colors of cubes to make a cube train to show the number pairs that make 8. 1–4. Complete the addition sentence to show a number pair for 8. Color the cube train to match the addition sentence in Exercise 4.

Lesson Check

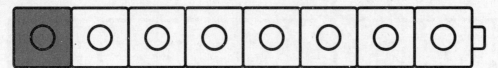

○ $8 = 1 + 7$　　　　○ $9 = 1 + 8$

○ $8 = 6 + 2$　　　　○ $9 = 7 + 2$

Spiral Review

2　　　　　3　　　　　4　　　　　5

○　　　　　○　　　　　○　　　　　○

○　　　　　○　　　　　○　　　　　○

DIRECTIONS　**1.** Which addition sentence shows a pair of numbers that matches the cube train? Mark beside your answer. **(Lesson 5.10)**　**2.** Mark under the number that is greater than the number of counters. **(Lesson 2.2)**　**3.** How many more counters would you place in the five frame to show a way to make 5? Mark under your answer. **(Lesson 1.7)**

Algebra • Number Pairs for 9

①

9 == ____ + ____

② **9** == ____ + ____

③ **9** == ____ + ____

④ **9** == ____ + ____

DIRECTIONS Use two colors of cubes to make a cube train to show the number pairs that make 9. **1–4.** Complete the addition sentence to show a number pair for 9. Color the cube train to match the addition sentence in Exercise 4.

Lesson Check

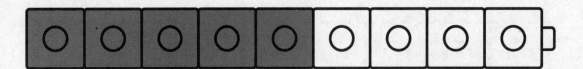

○ 5 = 4 + 1   ○ 8 = 5 + 3

○ 7 = 4 + 3   ○ 9 = 5 + 4

Spiral Review

8	7	6	5
○	○	○	○

2	3	4	5
○	○	○	○

DIRECTIONS **1.** Which addition sentence shows a pair of numbers that matches the cube train? Mark beside your answer. **(Lesson 5.11)** **2.** Count how many birds. Mark under your answer. **(Lesson 3.6)** **3.** Mark under the number that is less than the number of counters. **(Lesson 2.3)**

P102 one hundred two

Name _____

Algebra • Number Pairs for 10

1) $10 =$ ____ $+$ ____

2) $10 =$ ____ $+$ ____

3) $10 =$ ____ $+$ ____

4) $10 =$ ____ $+$ ____

DIRECTIONS Use two colors of cubes to build a cube train to show the number pairs that make 10. **1-4.** Complete the addition sentence to show a number pair for 10. Color the cube train to match the addition sentence in Exercise 4.

Lesson Check

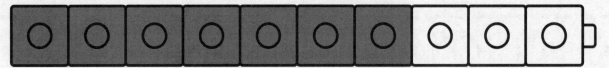

- ○ 7 = 3 + 4   ○ 9 = 6 + 3
- ○ 7 = 5 + 2   ○ 10 = 7 + 3

Spiral Review

- ○ 5, 3, 4, 1, 2   ○ 1, 2, 4, 5, 3
- ○ 1, 2, 3, 4, 5   ○ 3, 5, 1, 2, 4

○ [cube train]   ○ [cube train]

○ [cube train]   ○ [cube train]

DIRECTIONS **1.** Which addition sentence shows a pair of numbers that matches the cube train? Mark beside your answer. **(Lesson 5.12)** **2.** Which set of numbers is in order? Mark beside your answer. **(Lesson 1.8)** **3.** Which cube train shows a way to make 10? Mark beside your answer. **(Lesson 4.3)**

Chapter 5 Extra Practice

Lessons 5.1 - 5.3 (pp. 169–180) ·

1

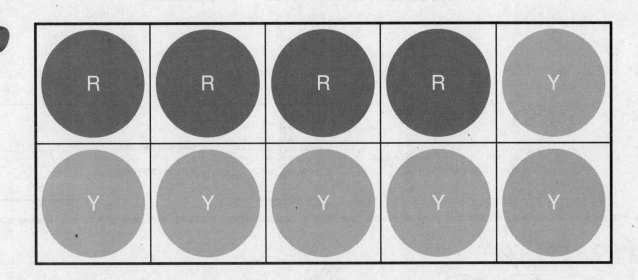

_____

- - - - -   **and**   - - - - -

_____

· ·

2

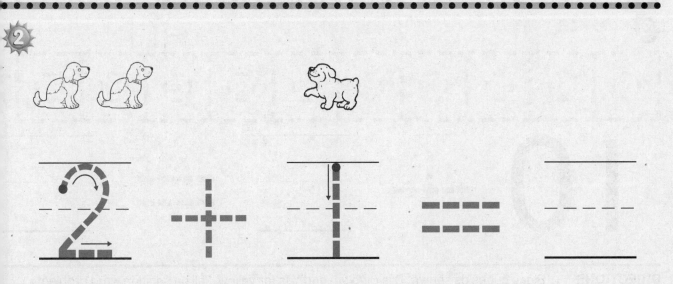

DIRECTIONS 1. Place counters in the ten frame as shown. R is for red, and Y is for yellow. How many of each color counter? Write the numbers. 2. Tell an addition word problem about the puppies. Trace the numbers and the symbols. Write the number that shows how many puppies there are now.

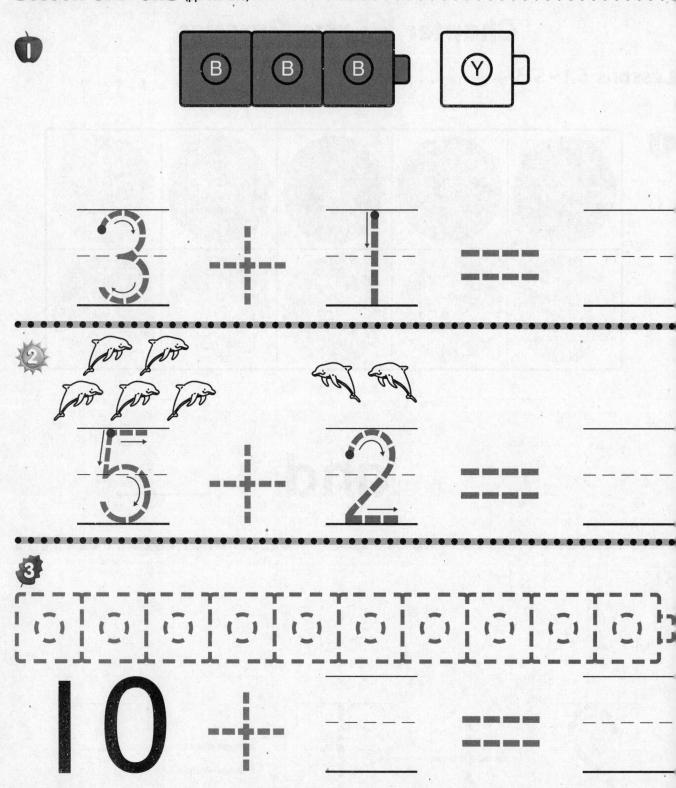

1.

$$3 + 1 = \underline{\quad}$$

2.

$$5 + 2 = \underline{\quad}$$

3.

$$10 + \underline{\quad} = \underline{\quad}$$

DIRECTIONS **1.** Place cubes as shown. B is for blue, and Y is for yellow. Tell an addition word problem. Model to show the cubes put together. Draw the cube train. Trace and write to complete the addition sentence. **2.** Tell an addition word problem. How many are in the set to start? How many are being added to the set? How many are there now? Write and trace to complete the addition sentence. **3.** Use two colors of cubes to build a cube train to show a number pair that makes 10. Complete the addition sentence to show a number pair for 10. Color the cube train to match the addition sentence.

School-Home Letter

Dear Family,

My class started Chapter 8 this week. In this chapter, I will learn how to show, count, and write numbers to 20 and beyond.

Love, _____

Vocabulary

twenty I ten and 10 ones

20

Home Activity

Make a set of number flash cards. Ask your child to lay out 20 cards to model what a set of 20 objects looks like. Then ask your child to place the number cards in the correct order from 1 to 20. Have your child point to each card and count forward from the number 1.

1	2	3	4	5	6	7	8	9	10

11	12	13	14	15	16	17	18	19	20

Literature

Look for these books at the library. Your child will enjoy these fun books while continuing to build counting skills.

20 Hungry Piggies by Trudy Harris. Millbrook Press, 2006.

Count! by Denise Fleming. Henry Holt and Co., 1995.

Carta
para la casa

Querida familia:

Mi clase comenzó el Capítulo 8 esta semana. En este capítulo, aprenderé cómo mostrar, contar y escribir números hasta el 20 y más allá. .

Con cariño, _____

Vocabulario

veinte una decena y 10 unidades

20

Actividad para la casa

Tome un conjunto de tarjetas nemotécnicas con números. Pídale a su hijo que separe 20 tarjetas para mostrar cómo es un conjunto de 20. Luego, pídale que ponga las tarjetas en el orden correcto del 1 al 20. Pídale a su hijo que señale cada carta y que cuente hacia delante desde el número 1.

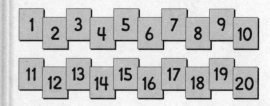

Literatura

Busque estos libros en la biblioteca. Su hijo disfrutará de estos libros divertidos mientras continua construir las habilidades de recuento.

20 Hungry Piggies por Trudy Harris. Millbrook Press, 2006.

Count! por Denise Fleming. Henry Holt and Co., 1995.

Name _____

Model and Count 20

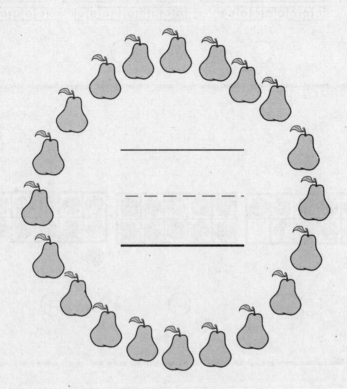

_____

- - - - - - - - - -

_____

_____

- - - - - - - - - -

_____

DIRECTIONS 1–2. Count and tell how many pieces of fruit. Write the
number. Tell a friend how you counted the fruit.

Lesson Check

1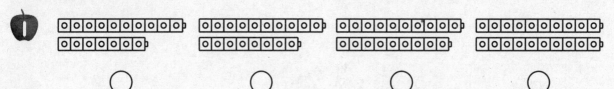

    ◯   ◯   ◯   ◯

Spiral Review

2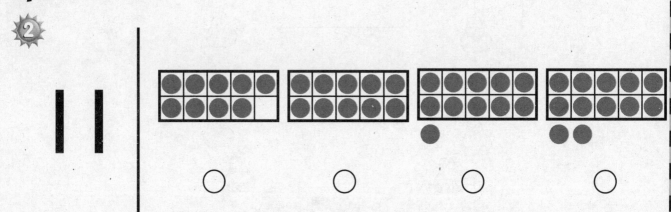

    ◯   ◯   ◯   ◯

3

$$1 \quad + \quad 4 \quad = \quad \underline{\qquad}$$

  2   3   4   5

  ◯   ◯   ◯   ◯

DIRECTIONS I. Which set of cubes models the number 20? Mark under your answer. **(Lesson 8.1)** **2.** Which set of counters shows the number I I? Mark under your answer. **(Lesson 7.1)** **3.** Which number completes the addition sentence about the sets of boats? Mark under your answer. **(Lesson 5.6)**

Read and Write to 20

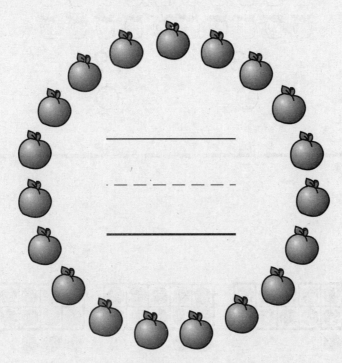

_____

- - - - - - - - - - -

_____

_____

- - - - - - - - - - -

_____

DIRECTIONS 1–2. Count and tell how many pieces of fruit. Write the number.

Lesson Check

17	18	19	20
○	○	○	○

- -

Spiral Review

14

         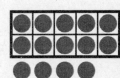

   ○

· ·

3

○ 1 + 3 = 4   ○ 2 + 1 = 3

○ 1 + 4 = 5   ○ 2 + 2 = 4

DIRECTIONS 1. Count and tell how many pieces of fruit. Mark under your answer. (Lesson 8.2) 2. Which set of counters shows the number 14? Mark under your answer. (Lesson 7.3) 3. Which addition sentence shows the cubes being put together? Mark beside your answer. (Lesson 5.4)

Name _____

Count and Order to 20

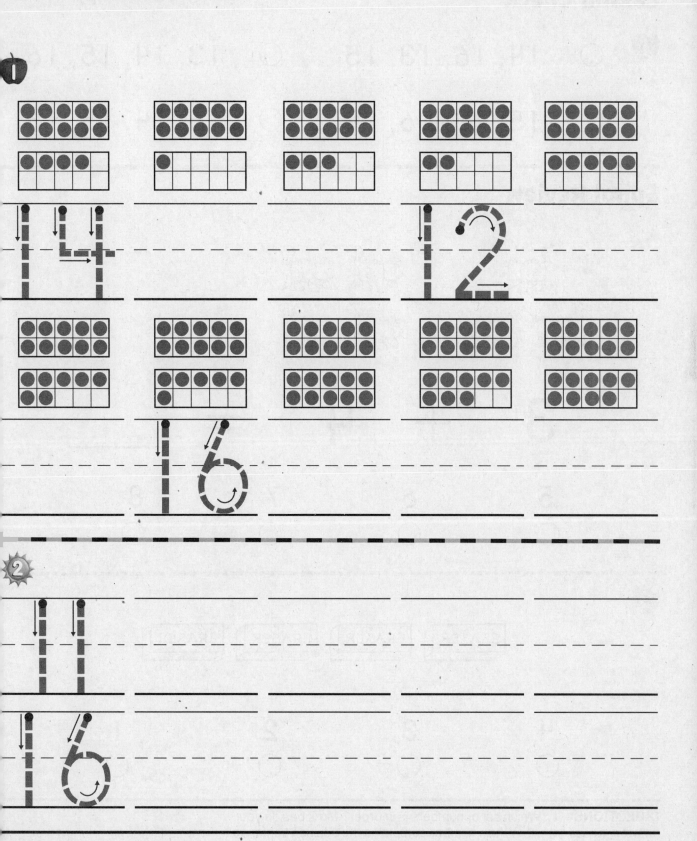

DIRECTIONS **1.** Count the dots in each set of ten frames. Trace or write the numbers. **2.** Trace and write those numbers in order.

Lesson Check

○ 14, 16, 13, 15   ○ 13, 14, 15, 16

○ 15, 13, 16, 14   ○ 16, 14, 15, 13

Spiral Review

$$3 + 4 = \underline{\hspace{1cm}}$$

5   6   7   8

○   ○   ○   ○

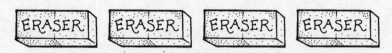

4   3   2   1

○   ○   ○   ○

DIRECTIONS **1.** Which set of numbers is in order? Mark beside your answer. (Lesson 8.3) **2.** Which number completes the addition sentence about the sets of cats? Mark under your answer. (Lesson 5.7) **3.** How many erasers are there? Mark under your answer. (Lesson 1.4)

**Problem Solving • Compare
Numbers to 20**

1

_____

- - - - - -

_____

_____

- - - - - -

_____

2

_____

- - - - - -

_____

_____

- - - - - -

_____

DIRECTIONS **1.** Teni has 16 berries. She has a number of berries two greater than Marta. Use cubes to model the sets of berries. Compare the sets. Which set is larger? Draw the cubes. Write how many in each set. Circle the greater number. Tell a friend how you compared the numbers. **2.** Ben has 18 pears. Sophia has a number of pears two less than Ben. Use cubes to model the sets of pears. Compare the sets. Which set is smaller? Draw the cubes. Write how many in each set. Circle the number that is less. Tell a friend how you compared the numbers.

Lesson Check

1

○   ○   ○   ○

Spiral Review

2

16

○   ○   ○   ○

3

1   2   3   4

○   ○   ○   ○

DIRECTIONS 1. Compare the sets. Which set has a number of cubes two less than 20? Mark under your answer. **(Lesson 8.4)** 2. Which set of counters shows the number 16? Mark under your answer. **(Lesson 7.7)** 3. Mark under the number that is greater than the number of counters. **(Lesson 2.2)**

Count to 50 by Ones

1	2	3	4	5	6	7	8	9	10
11	12	13	14	15	16	17	18	19	20
21	22	23	24	25	26	27	28	29	30
31	32	33	34	35	36	37	38	39	40
41	42	43	44	45	46	47	48	49	50

DIRECTIONS   1. Look away and point to any number. Circle that number. Count forward from that number. Draw a line under the number 50.

Lesson Check

1	2	3	4	5	6	7	8	9	10
11	12	13	14	15	16	17	18	19	20
21	22	23	24	25	26	27	28	29	30

   20   21   22   23

   ○   ○   ○   ○

Spiral Review

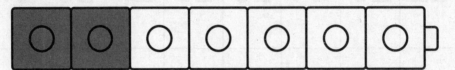

$6 = 5 + 1$	$5 = 2 + 3$	$6 = 2 + 4$	$7 = 2 + 5$
○	○	○	○

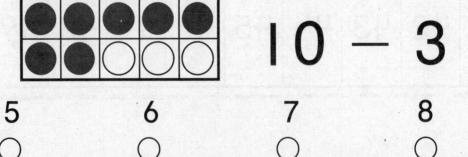

$$10 - 3$$

  5   6   7   8

  ○   ○   ○   ○

DIRECTIONS 1. Begin with 1 and count forward to 20. What is the next number? Mark under your answer. **(Lesson 8.5) 2.** Which addition sentence shows a numbers pair that matches the cube train? Mark under your answer. **(Lesson 5.9) 3.** Shelley has 10 counters. Three of her counters are white. The rest of her counters are gray. How many are gray? Mark under your answer. **(Lesson 6.2)**

School-Home Letter

Dear Family,

My class started Chapter 9 this week. In this chapter, I will learn how to identify, name, and describe two-dimensional shapes.

Love, _____

Vocabulary

curve a line that is rounded

vertex the point where two sides of a two-dimensional shape meet

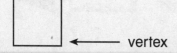

Home Activity

Spread out a group of household objects. Have your child point out the objects that look like circles, squares, and triangles.

Literature

Look for these books at the library. The pictures will capture your child's imagination.

Shapes, Shapes, Shapes by Tana Hoban. Greenwillow, 1996.

Color Farm by Lois Ehlert. HarperCollins, 1990.

Carta para la casa

Querida familia:

Mi clase comenzó el Capítulo 9 esta semana. En este capítulo, aprenderé cómo identi car, nombrar y describir guras bidimensionales.

Con cariño, _____

Vocabulario

curva una línea que no es recta

vértice el punto en donde se encuentran dos lados de una figura bidimensional

← vértice

Actividad para la casa

Dé a su hijo varios objetos que encuentre en la casa y pídale que señale los que se parezcan a los cuadrados, círculos y triángulos.

Literatura

Busque este libro en la biblioteca. Las ilustraciones estimularán la imaginación de su hijo.

Shapes, Shapes, Shapes
por Tana Hoban.
Greenwillow, 1996.

Color Farm
por Lois Ehlert.
HarperCollins, 1990.

Identify and Name Circles

DIRECTIONS   1. Color the circles in the picture.

Lesson Check

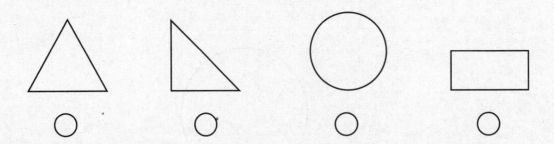

○   ○   ○   ○

Spiral Review

17, 18, 16 | 18, 16, 17 | 16, 17, 18 | 16, 18, 17

○   ○   ○   ○

$$5 + 3 = \underline{\hspace{2cm}}$$

5   6   7   8

○   ○   ○   ○

DIRECTIONS   **1.** Which shape is a circle? Mark under your answer.
(Lesson 9.1)   **2.** Which set of numbers is in order? Mark under your
answer. (Lesson 8.3)   **3.** Which number completes the addition sentence
about the sets of cats? Mark under your answer. (Lesson 5.7)

Describe Circles

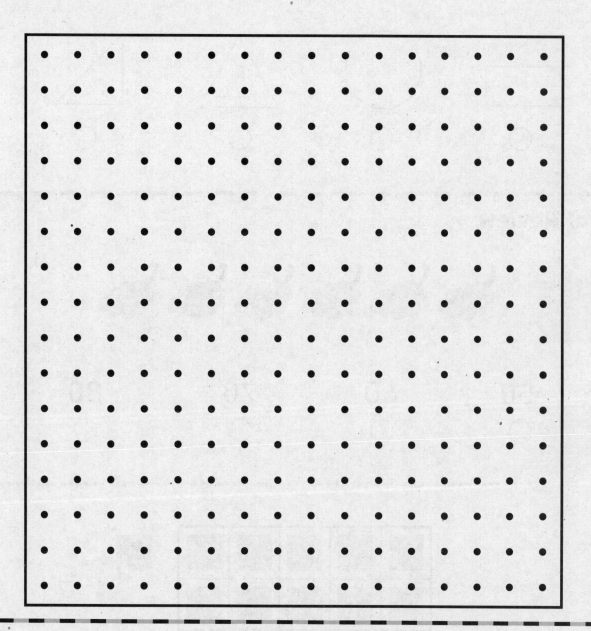

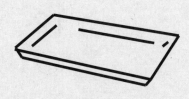  

DIRECTIONS **1.** Use a pencil to hold one end of a large paper clip on one of the dots in the center. Place another pencil in the other end of the paper clip. Move the pencil around to draw a circle. **2.** Color the object that is shaped like a circle.

Chapter 9

Lesson Check

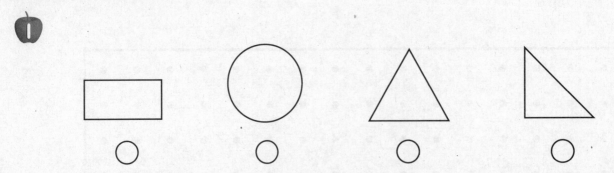

○   ○   ○   ○

Spiral Review

50   60   70   80
○   ○   ○   ○

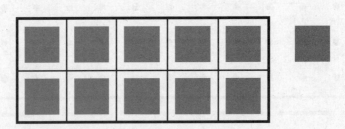

9   10   11   12
○   ○   ○   ○

DIRECTIONS **1.** Which shape has a curve? Mark under your answer.
(Lesson 9.2) **2.** Point to each set of 10 as you count by tens. Mark under the
number that shows how many grapes there are. **(Lesson 8.8)** **3.** How many
tiles are there? Mark under your answer. **(Lesson 7.2)**

Identify and Name Squares

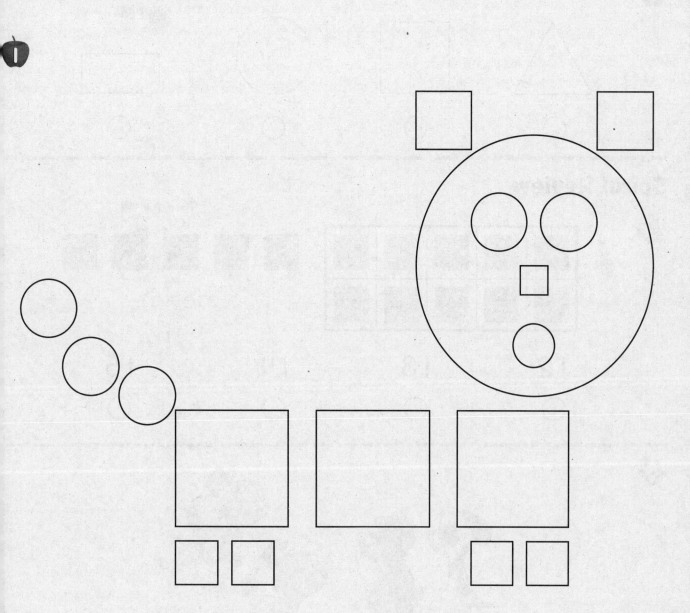

Lesson Check

1

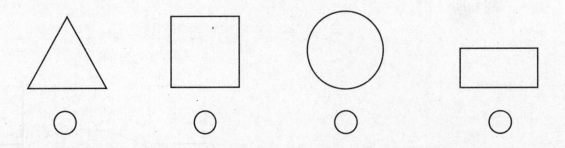

○   ○   ○   ○

Spiral Review

2

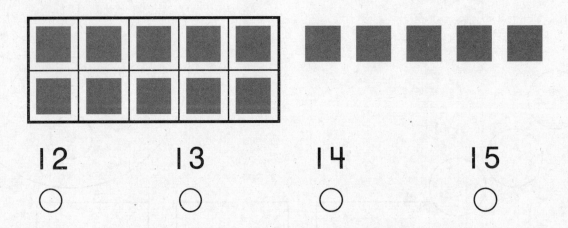

12   13   14   15

○   ○   ○   ○

3

2   3   4   5

○   ○   ○   ○

DIRECTIONS   1. Which shape is a square? Mark under your answer.
(Lesson 9.3)   **2.** How many tiles are there? Mark under your answer.   (Lesson 7.6)
3. How many puppies are there in all? Mark under your answer.   (Lesson 5.3)

Name _____

Describe Squares

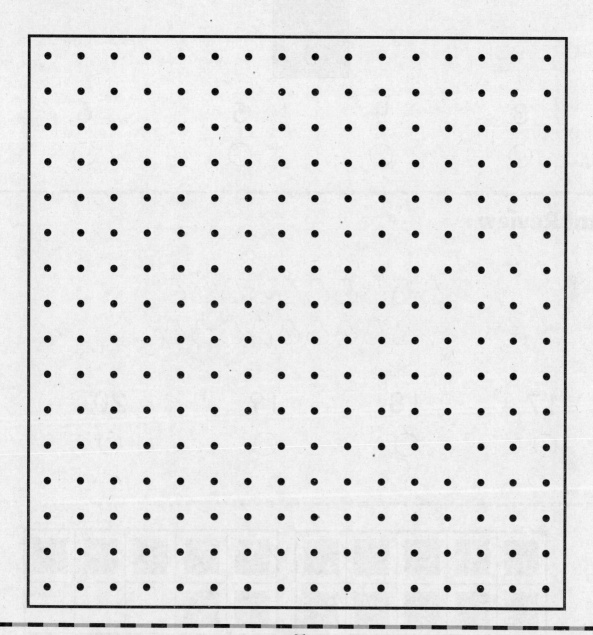

2 _____ **3** _____

_____ _____

_____ **vertices** _____ **sides**

DIRECTIONS I. Draw and color a square. **2.** Place a counter on each corner, or vertex, of the square that you drew. Write how many corners, or vertices. **3.** Trace around the sides of the square that you drew. Write how many sides.

Lesson Check

3   4   5   6

○   ○   ○   ○

Spiral Review

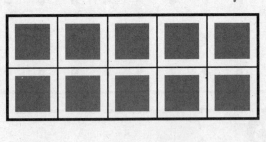

17   18   19   20

○   ○   ○   ○

14   15   16   17

○   ○   ○   ○

DIRECTIONS **1.** How many vertices does the square have? Mark under your answer. **(Lesson 9.4)** **2.** Count and tell how many pieces of fruit. Mark under your answer. **(Lesson 8.2)** **3.** How many tiles are there? Mark under your answer. **(Lesson 7.8)**

Identify and Name Triangles

 1

 2

DIRECTIONS 1–2. Color the triangles in the picture.

Lesson Check

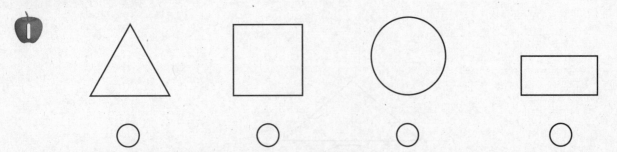

○　　　　　○　　　　　○　　　　　○

Spiral Review

1	2	3	4	5	6	7	8	9	10
11	12	13	14	15	16	17	18	19	20
21	22	23	24	25	26	27	28	29	30

24　　　　25　　　　26　　　　27

○　　　　　○　　　　　○　　　　　○

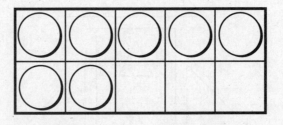

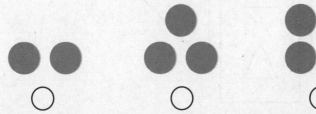

○　　　　　○　　　　　○　　　　　○

DIRECTIONS **1.** Which shape is a triangle? Mark under your answer.
(Lesson 9.5) **2.** Begin with 1 and count forward to 24. What is the next number?
Mark under your answer. **(Lesson 8.5)** **3.** How many more counters would you
place to model a way to make 10? Mark under your answer. **(Lesson 4.1)**

Describe Triangles

② _____ ③ _____

- - - - -   - - - - -

_____ **vertices** _____ **sides**

DIRECTIONS 1. Draw and color a triangle. 2. Place a counter on each corner, or vertex, of the triangle that you drew. Write how many corners, or vertices. 3. Trace around the sides of the triangle that you drew. Write how many sides.

Lesson Check

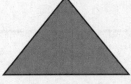

1	2	3	4
○	○	○	○

Spiral Review

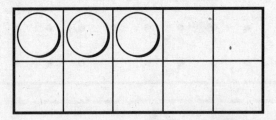

$$5 - 2 = \underline{\hspace{2cm}}$$

2	3	4	5
○	○	○	○

○   ○   ○   ○

DIRECTIONS **I.** How many sides does the triangle have? Mark under your answer. (Lesson 9.6)   **2.** Which number shows how many kittens are left? Mark under your answer. (Lesson 6.3)   **3.** How many more counters would you place to model a way to make 7? Mark under your answer. (Lesson 3.3)

Identify and Name Rectangles

Lesson Check

1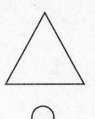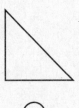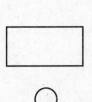

    ◯   ◯   ◯   ◯

Spiral Review

2

1	2	3	4	5	6	7	8	9	10
11	12	13	14	15	16	17	18	19	20
21	22	23	24	25	26	27	28	29	30

  10   15   27   30

  ◯   ◯   ◯   ◯

3

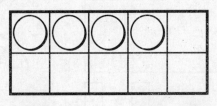

   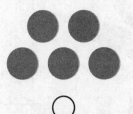

  ◯   ◯   ◯   ◯

DIRECTIONS 1. Which shape is a rectangle? Mark under your answer.
(Lesson 9.7) 2. Count by tens as you point to the numbers in the shaded boxes. Start with the number 10. What number do you end with? Mark under your answer. (Lesson 8.7) 3. How many more counters would you place to model a way to make 6? Mark under your answer. (Lesson 3.1)

Describe Rectangles

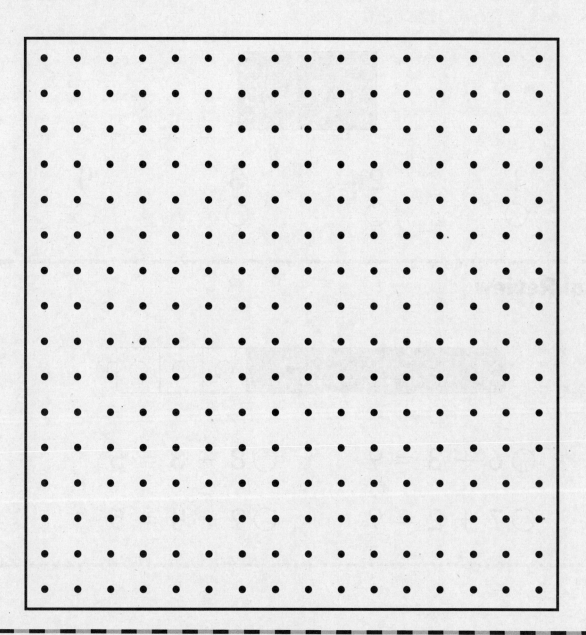

 vertices | **sides**

2 _____

- - - - -

_____ **vertices**

3 _____

- - - - -

_____ **sides**

DIRECTIONS 1. Draw and color a rectangle. 2. Place a counter on each corner, or vertex, of the rectangle that you drew. Write how many corners, or vertices. 3. Trace around the sides of the rectangle that you drew. Write how many sides.

Lesson Check

1	2	3	4
○	○	○	○

Spiral Review

○ 6 + 3 = 9   ○ 8 − 3 = 5

○ 7 + 2 = 9   ○ 9 − 4 = 5

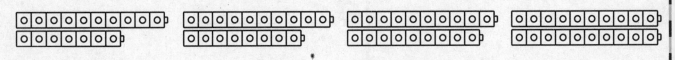

  ○   ○   ○   ○

DIRECTIONS **1.** How many sides does the rectangle have? Mark under your answer. **(Lesson 9.8)** **2.** Mark beside the number sentence that matches the picture. **(Lesson 6.7)** **3.** Compare the sets. Which set has a number of cubes two greater than 18? Mark under your answer. **(Lesson 8.4)**

Identify and Name Hexagons

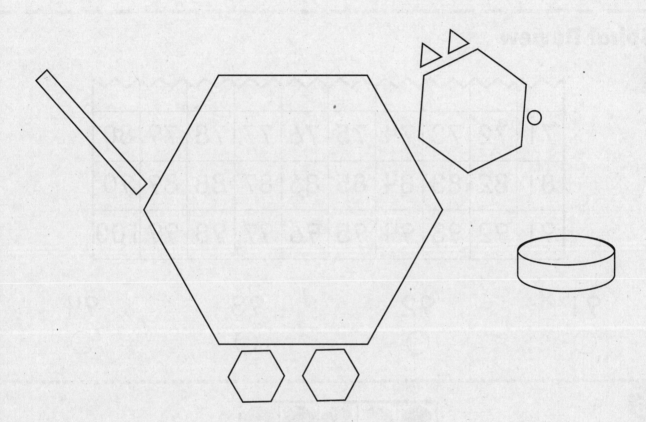

DIRECTIONS 1. Color the hexagons in the picture.

Lesson Check

□   ▭   △   ⬡

○   ○   ○   ○

Spiral Review

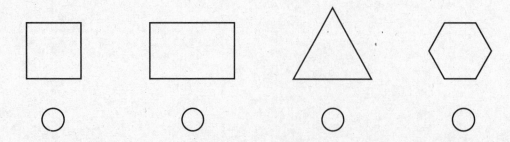

71	72	73	74	75	76	77	78	79	80
81	82	83	84	85	86	87	88	89	90
91	92	93	94	95	96	97	98	99	100

91   92   93   94
○   ○   ○   ○

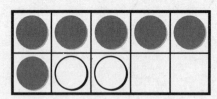

5 + 1   6 + 2   7 + 3   8 + 1
○   ○   ○   ○

DIRECTIONS **1.** Which shape is a hexagon? Mark under your answer.
(Lesson 9.9) **2.** Begin with 81 and count forward to 90. What is the next
number? Mark under your answer. **(Lesson 8.6)** **3.** Which numbers show the
sets that are put together? Mark under your answer. **(Lesson 5.2)**

Name _____

Describe Hexagons

2 _____

    - - - - -

    _____ **vertices**

3 _____

    - - - - -

    _____ **sides**

DIRECTIONS 1. Draw and color a hexagon. 2. Place a counter on each corner, or vertex, of the hexagon that you drew. Write how many corners, or vertices. 3. Trace around the sides of the hexagon that you drew. Write how many sides.

Lesson Check

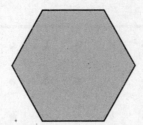

7	6	5	4
○	○	○	○

Spiral Review

○ 6 = 3 + 3   ○ 8 = 6 + 2

○ 7 = 6 + 1   ○ 9 = 6 + 3

6

4	7	5	6
○	○	○	○

DIRECTIONS **1.** How many sides does the hexagon have? Mark under your answer. (Lesson 9.10) **2.** Which addition sentence shows a number pair that matches the cube train? Mark beside your answer. (Lesson 5.11) **3.** Which number is greater than 6? Mark under your answer. (Lesson 4.7)

Algebra • Compare
Two-Dimensional Shapes

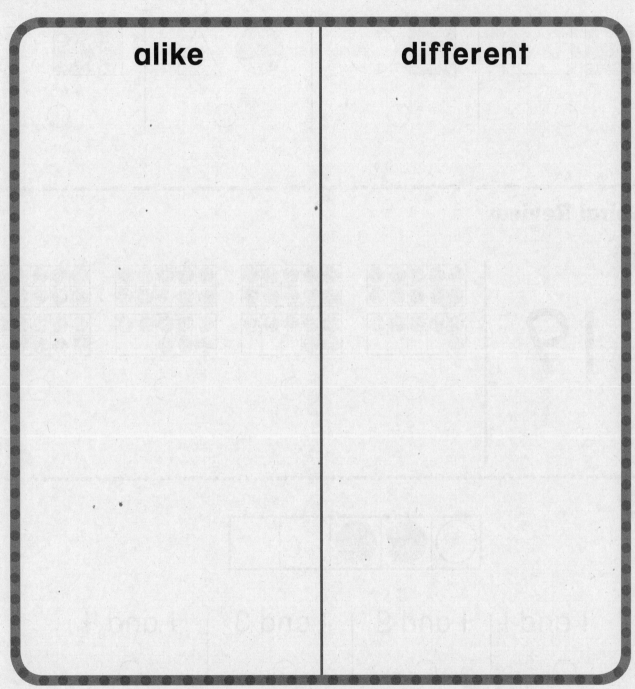

alike	different

DIRECTIONS I. Place two-dimensional shapes on the page. Sort the shapes by the number of sides. Draw the shapes on the sorting mat. Use the words *alike* and *different* to tell how you sorted the shapes.

Lesson Check

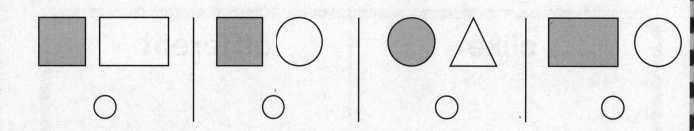

Spiral Review

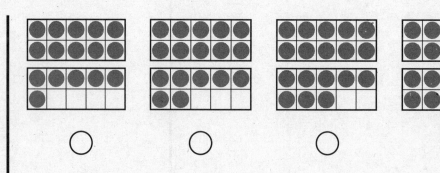

19

| I and I | I and 2 | I and 3 | I and 4 |

○   ○   ○   ○

DIRECTIONS 1. Which two shapes are alike in some way? Mark under your answer. (Lesson 9.11) 2. Which set of counters shows the number 19? Mark under your answer. (Lesson 7.9) 3. Which shows the gray counters being added to the five frame? Mark under your answer. (Lesson 5.1)

Problem Solving • Draw to Join Shapes

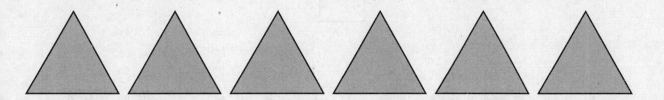

1

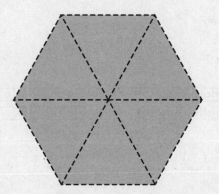

2

DIRECTIONS 1. Place triangles on the page as shown. How can you join all of the triangles to make a hexagon? Trace around the triangles to draw the hexagon. 2. How can you join some of the triangles to make a larger triangle? Trace around the triangles to draw the larger triangle.

Lesson Check

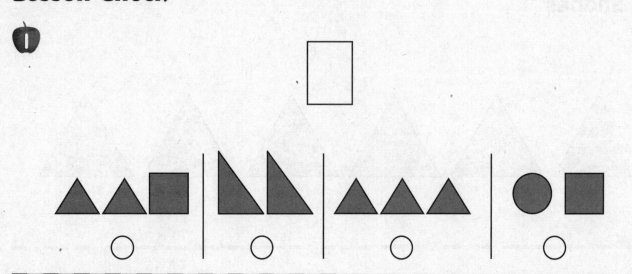

Spiral Review

②

⃝   ⃝   ⃝   ⃝

③

5   6   7   8

⃝   ⃝   ⃝   ⃝

DIRECTIONS 1. Which shapes could you join to make the rectangle above? Mark under your answer. **(Lesson 9.12) 2.** Which set of cubes models the number 20? Mark under your answer. **(Lesson 8.1) 3.** Mark under the number that is less than the number of spoons. **(Lesson 4.6)**

Chapter 9 Extra Practice

Lessons 9.1 - 9.6 (pp. 357–379) •

1

 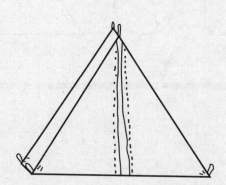 

• •

2

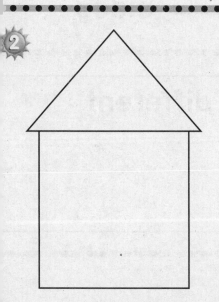 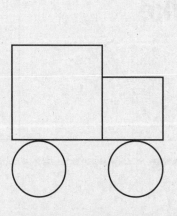 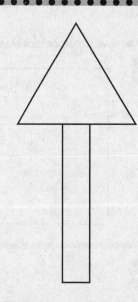

• •

3

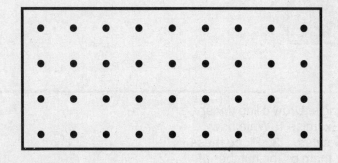

DIRECTIONS   I. Color the object that is shaped like a circle.   **2.** Use red to color the squares in the picture. Use green to color the triangles.   **3.** Draw and color a triangle.

1

2

_____   _____

- - - - -   - - - -

_____ **vertices**   _____ **sides**

3

alike	**different**

4

DIRECTIONS   1. Mark an X on the rectangle. Draw a line under the hexagon.   **2.** Look at the hexagon in Exercise 1. Write how many corners, or vertices. Write how many sides.   **3.** Place these two-dimensional shapes on the page. Sort them by the number of vertices as shown. Trace the shapes. Color the shapes that have three vertices.   **4.** Place two squares on the page as shown. How can you join the squares to make a rectangle? Trace around the squares to draw the rectangle.

School-Home Letter

Dear Family,

My class started Chapter 10 this week. In this chapter, I will learn how identifying and describing shapes can help me sort them.

Love, _____

Vocabulary

sphere a three-dimensional shape that is round
A ball is an example of a sphere.

cylinder a three-dimensional shape with a curved surface and two flat surfaces

Home Activity

Take a walk around your neighborhood with your child. Ask your child to point out objects that are shaped like three-dimensional shapes, such as spheres, cubes, cylinders, and cones.

Recycle

Literature

Look for these books at the library. The pictures will help your child understand how shapes are a part of everyday life.

What in the World Is a Sphere?
by Anders Hanson. SandCastle, 2007.

Cubes, Cones, Cylinders, & Spheres
by Tana Hoban. Greenwillow Books, 2000.

Carta para la casa

Querida familia:

Mi clase comenzó el Capítulo 10 esta semana. En este capítulo, aprenderé cómo identificar y describir figuras puede ayudarme a clasificarlas.

Con cariño, _____

Vocabulario

esfera una figura tridimensional redonda

Una pelota es un ejemplo de esfera.

cilindro una figura tridimensional con una superficie curva y dos superficies planas

Actividad para la casa

Salga a caminar por el barrio junto a su hijo. Pídale que señale objetos que tengan formas tridimensionales, tales como esferas, cubos, cilindros y conos.

Recycle

Literatura

Busque estos libros en la biblioteca. Los dibujos ayudarán a que su hijo comprenda cómo las figuras forman parte de la vida diaria.

What in the World Is a Sphere? por Anders Hanson. SandCastle, 2007.

Cubes, Cones, Cylinders & Spheres por Tana Hoban. Greenwillow Books, 2000

Three-Dimensional Shapes

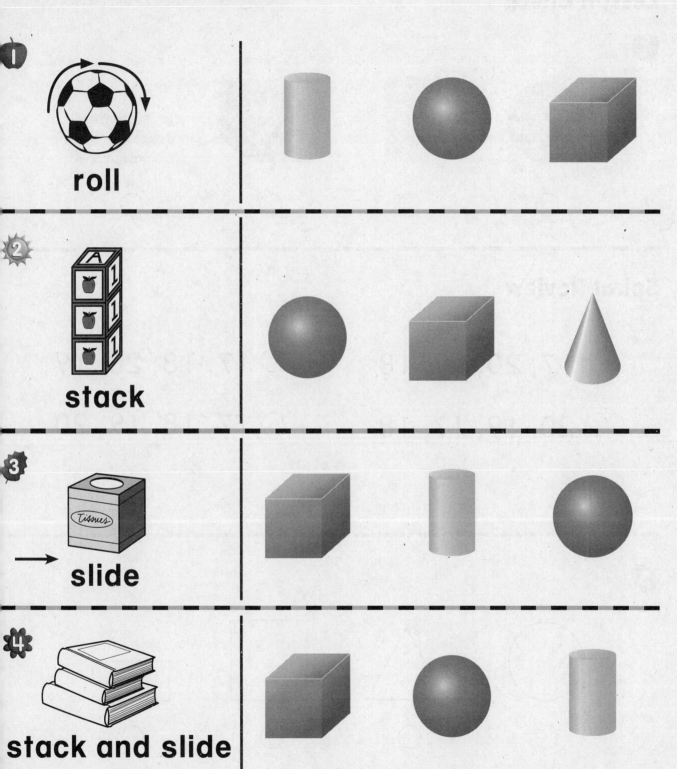

1 roll

2 stack

3 slide

4 stack and slide

DIRECTIONS 1. Which shape does not roll? Mark an X on that shape. 2. Which shapes do not stack? Mark an X on those shapes. 3. Which shape does not slide? Mark an X on that shape. 4. Which shape does not stack and slide? Mark an X on that shape.

Lesson Check

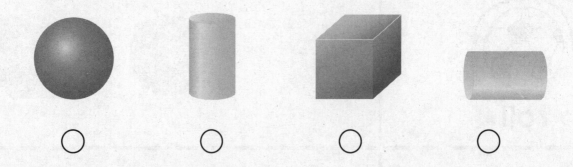

○   ○   ○   ○

Spiral Review

○ 17, 20, 19, 18   ○ 17, 18, 20, 19

○ 20, 19, 17, 18   ○ 17, 18, 19, 20

· ·

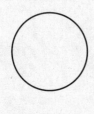   

○   ○   ○   ○

DIRECTIONS 1. Which shape does not roll? Mark under your answer. (Lesson 10.1)   2. Which set of numbers is in order? Mark beside your answer. (Lesson 8.3)   3. Which shape has a curve? Mark under your answer. (Lesson 9.2)

Identify, Name, and Describe Spheres

DIRECTIONS 1. Identify the objects that are shaped like a sphere. Mark an X on those objects.

Lesson Check

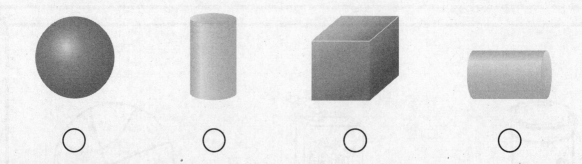

○ ○ ○ ○

Spiral Review

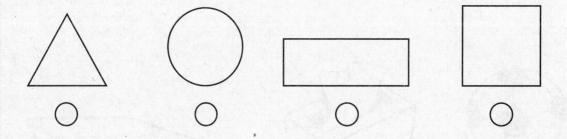

○ ○ ○ ○

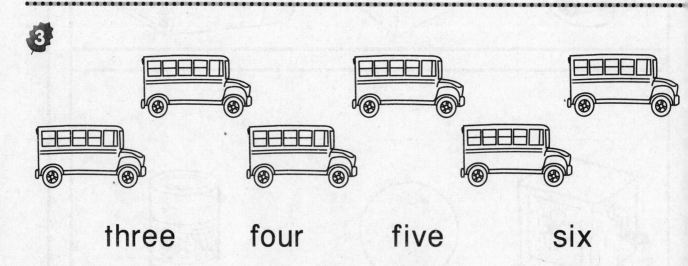

| three | four | five | six |

○ ○ ○ ○

DIRECTIONS   1. Which shape is a sphere? Mark under your answer. (Lesson 10.2)
2. Which shape is a square? Mark under your answer. (Lesson 9.3)
3. How many school buses are there? Mark under your answer. (Lesson 3.2)

Identify, Name, and Describe Cubes

DIRECTIONS 1. Identify the objects that are shaped like a cube. Mark an X on those objects.

Lesson Check

1

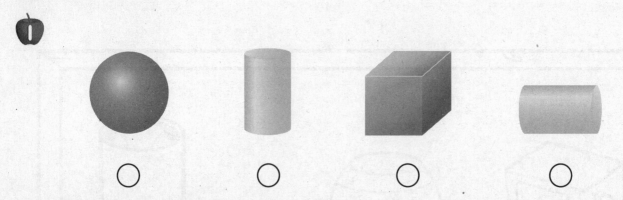

○   ○   ○   ○

Spiral Review

2

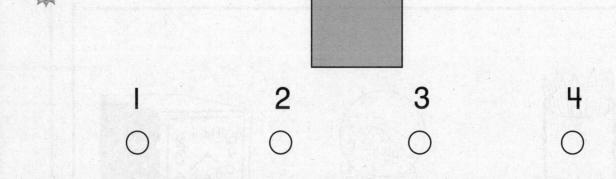

| 1 | 2 | 3 | 4 |

○   ○   ○   ○

3

71	72	73	74	75	76	77	78	79	80
81	82	83	84	85	86	87	88	89	90
91	92	93	94	95	96	97	98	99	100

89   91   98   100

○   ○   ○   ○

DIRECTIONS 1. Which shape is a cube? Mark under your answer. **(Lesson 10.3)**
2. How many sides does the square have? Mark under your answer. **(Lesson 9.4)**
3. Begin with 81 and count forward to 90. What is the next number? Mark under your answer. **(Lesson 8.6)**

Identify, Name, and Describe Cylinders

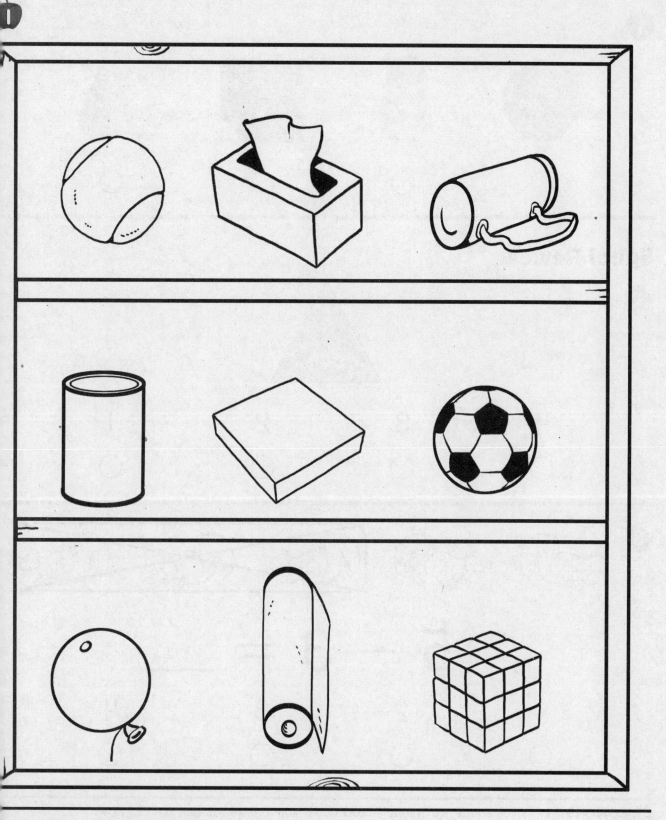

DIRECTIONS 1. Identify the objects that are shaped like a cylinder. Mark an X on those objects.

two hundred five **P205**

Lesson Check

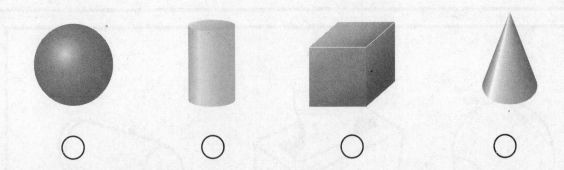

○   ○   ○   ○

- -

Spiral Review

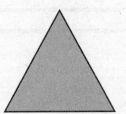

4   3   2   1

○   ○   ○   ○

· ·

$$5 - 3 = \underline{\quad}$$

1   2   3   4

○   ○   ○   ○

DIRECTIONS I. Which shape is a cylinder? Mark under your answer. **(Lesson 10.4)** **2.** How many vertices does the triangle have? Mark under your answer. **(Lesson 9.6)** **3.** Which number completes the subtraction sentence about the set of squirrels? Mark under your answer. **(Lesson 6.5)**

Identify, Name, and Describe Cones

DIRECTIONS 1. Identify the objects that are shaped like a cone. Mark an X on those objects.

Chapter 10   two hundred seven P207

Lesson Check

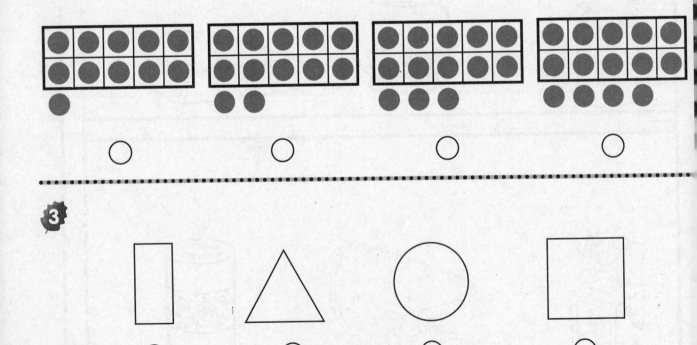

○   ○   ○   ○

Spiral Review

2

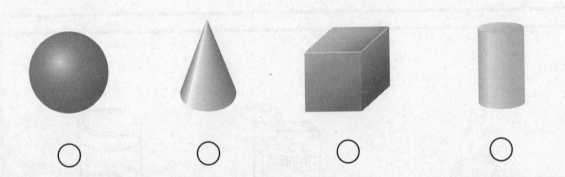

13

○   ○   ○   ○

3

○   ○   ○   ○

DIRECTIONS   1. Which shape is a cone? Mark under your answer. (Lesson 10.5)
2. Which set of counters shows the number 13? Mark under your answer. (Lesson 7.3)
3. Which shape is a circle? Mark under your answer. (Lesson 9.1)

Name _____

Problem Solving • Two- and Three-Dimensional Shapes

DIRECTIONS 1. Identify the two-dimensional or flat shapes. Use red to color the flat shapes. Identify the three-dimensional or solid shapes. Use blue to color the solid shapes.

Lesson Check

1

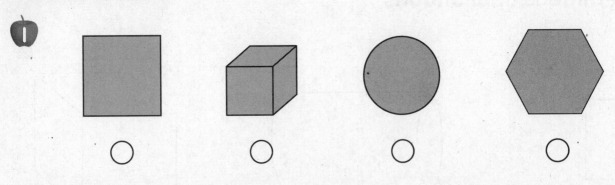

○   ○   ○   ○

Spiral Review

2

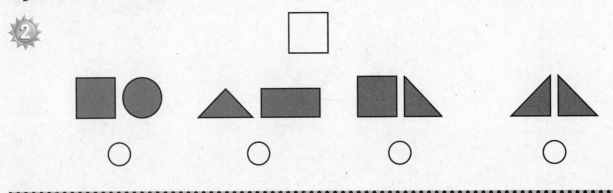

○   ○   ○   ○

3

1	2	3	4	5	6	7	8	9	10
11	12	13	14	15	16	17	18	19	20
21	22	23	24	25	26	27	28	29	30

18   19   20   21

○   ○   ○   ○

DIRECTIONS 1. Which is a three-dimensional or solid shape? Mark under your answer. **(Lesson 10.6)** **2.** Which shapes could you join to make the square above? Mark under your answer. **(Lesson 9.12)** **3.** Begin with 1 and count forward to 19. What is the next number? Mark under your answer. **(Lesson 8.5)**

Above and Below

CAT
TOYS

DIRECTIONS **I.** Mark an X on the object that is shaped like a sphere below
the table. Circle the object that is shaped like a cube above the table.

Chapter 10   two hundred eleven **P2 I I**

Lesson Check

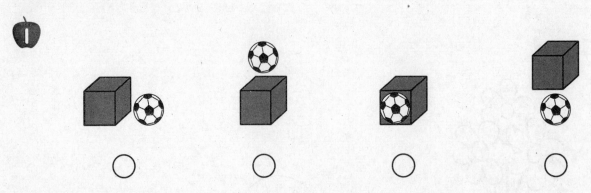

○   ○   ○   ○

Spiral Review

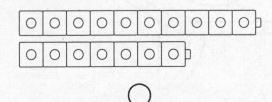

○

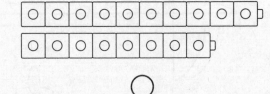

○

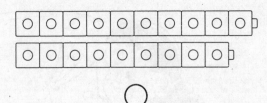

○

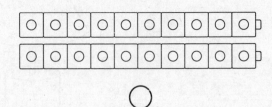

○

· ·

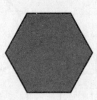

5   **6**   **7**   **8**

○   ○   ○   ○

DIRECTIONS I. Which picture shows that the object shaped like a sphere is above the box? Mark under your answer. (Lesson 10.7) 2. Which set of cubes models the number 20? Mark under your answer. (Lesson 8.1) 3. How many vertices does the hexagon have? Mark under your answer. (Lesson 9.10)

Beside and Next To

DIRECTIONS **1.** Mark an X on the object shaped like a cylinder that is next to the object shaped like a sphere. Circle the object shaped like a cone that is beside the object shaped like a cube. Use the words *next to* and *beside* to name the position of other shapes.

Lesson Check

 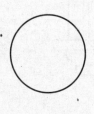  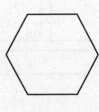

Spiral Review

 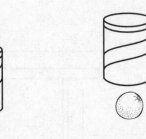 

   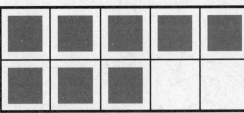

16   17   18   19

○   ○   ○   ○

DIRECTIONS 1. Which picture shows an object shaped like a sphere is beside an object shaped like a cylinder? Mark under your answer. (Lesson 10.8) 2. Which shape is a hexagon? Mark under your answer. (Lesson 9.9) 3. How many tiles are there? Mark under your answer. (Lesson 7.10)

n Front Of and Behind

DIRECTIONS **1.** Mark an X on the object shaped like a cylinder that is behind the object shaped like a cone. Draw a circle around the object shaped like a cylinder that is in front of the object shaped like a cube. Use the words in front of and behind to name the position of other shapes.

Chapter 10 two hundred fifteen **P215**

Lesson Check

 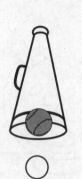  

○　　　○　　　○　　　○

Spiral Review

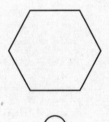  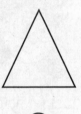 

○　　　○　　　○　　　○

···

3 and 0　　3 and 1　　3 and 2　　3 and 3

DIRECTIONS **1.** Which picture shows an object shaped like a sphere in front of an object shaped like a cone? Mark under your answer. **(Lesson 10.9)**
2. Which shape is a triangle? Mark under your answer. **(Lesson 9.5)**
3. Which shows the gray counters being added to the five frame? Mark under your answer. **(Lesson 5.1)**

Chapter 10 Extra Practice

Lessons 10.1 – 10.5 (pp. 413–432) ·

1

· ·

2
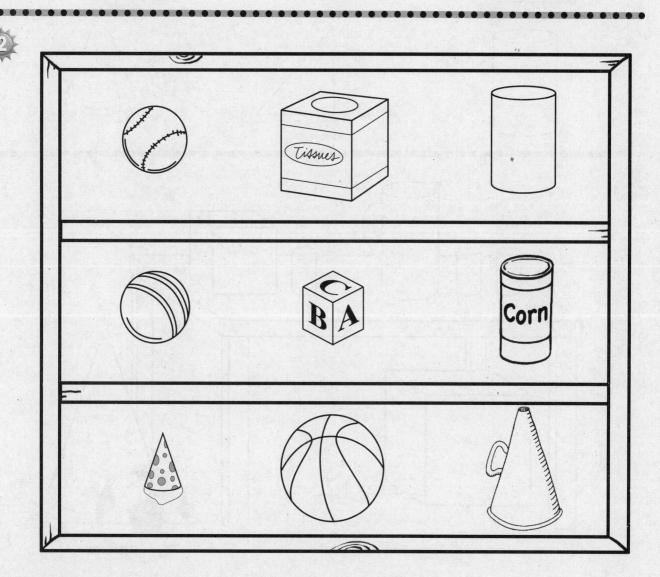

DIRECTIONS 1. Which shapes do not stack? Mark an X on those shapes.
2. Identify the objects that are shaped like a sphere. Color those objects. Identify
the objects that are shaped like a cube. Circle those objects. Identify the objects
that are shaped like a cone. Mark an X on those objects. Identify the objects that
are shaped like a cylinder. Draw a line under those objects.

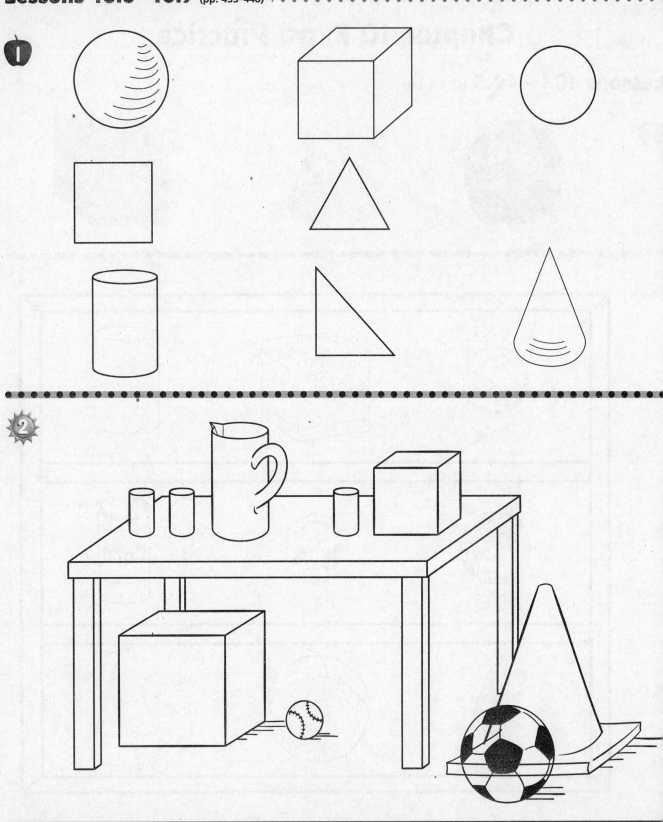

DIRECTIONS I. Identify the two-dimensional or flat shapes. Use red to color the flat shapes. Identify the three-dimensional or solid shapes. Use blue to color the solid shapes. 2. Mark an X on the object shaped like a cube that is below the table. Draw a circle around the object shaped like a cylinder that is beside the object shaped like a cube. Color the object shaped like a sphere that is in front of the object shaped like a cone.

Chapter 12 Extra Practice

Lessons 12.1–12.3 (pp. 235–240)

rectangle	triangle

2 | rectangle | triangle |

_____

- - - - - -

_____

DIRECTIONS 1. Place a red triangle, red rectangle, green triangle, red triangle, and blue rectangle at the top of the page as shown. Sort and classify the shapes by the category of shape. Draw and color the shapes in each category. Look at the categories in Exercise 1. Count how many in each category. **2.** Circle the category that has two shapes. Write the number.

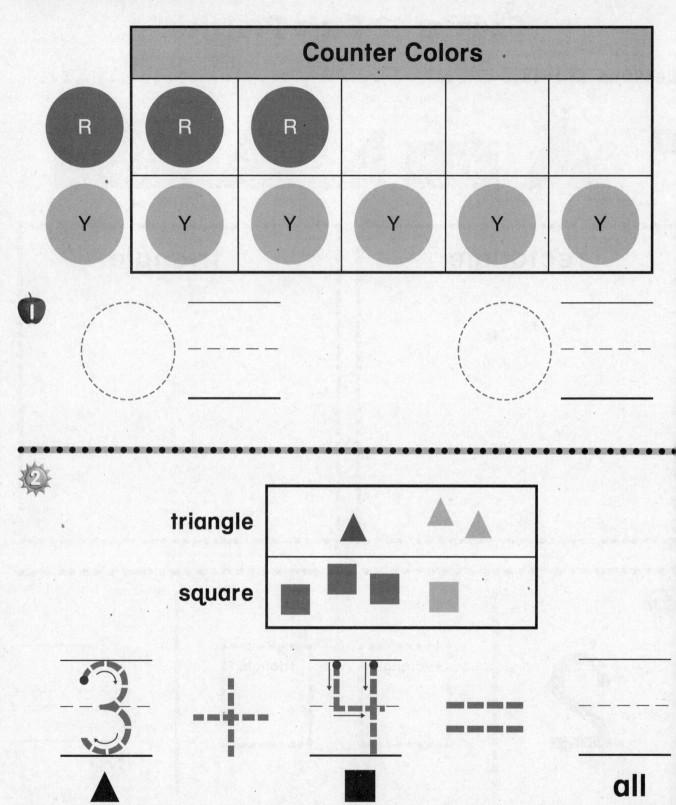

Counter Colors

①

②

triangle

square

$3 + 4 =$ ____

▲   ■   **all**

DIRECTIONS 1. Color the counters to show the categories. R is for red, and Y is for yellow. How many counters are in each category? Write the numbers. 2. Look at the sorting mat. How are the shapes sorted? How many triangles are shown? How many squares are shown? Add the two sets. Trace and write to complete the addition sentence.

Add One

1 $1 + 1 = 2$

2 $2 + \underline{\quad} = \underline{\quad}$

3 $3 + \underline{\quad} = \underline{\quad}$

DIRECTIONS 1. Place cubes as shown above the numbers.
Trace the cubes. Trace to complete the addition sentence.
2–3. Use cubes to show the number. Draw the cubes.
Show and draw one more cube. Complete the addition sentence.

4

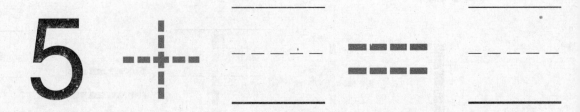

5

6

DIRECTIONS   4–6. Use cubes to show the number. Draw the cubes. Show and draw one more cube. Complete the addition sentence.

HOME ACTIVITY · Show your child a set of one to nine pennies. Have him or her use pennies to show how to add one to the set. Then have him or her tell how many in all.

Add Two

1

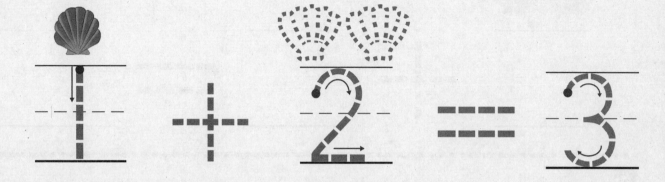

· ·

2

 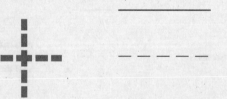 

· ·

3

 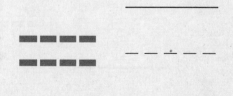

· ·

DIRECTIONS 1. Count how many shells in the first group. Trace the two shells. Trace to complete the addition sentence. **2–3.** Count how many shells. Write the number. Draw two more shells. Complete the addition sentence.

4

_____   _ _ _ _ _ _ $+$ _____   _ _ _ _ _ _ $=$ _____   _ _ _ _ _ _

5

_____   _ _ _ _ _ _ $+$ _____   _ _ _ _ _ _ $=$ _____   _ _ _ _ _ _

6

_____   _ _ _ _ _ _ $+$ _____   _ _ _ _ _ _ $=$ _____   _ _ _ _ _ _

DIRECTIONS 4–6. Count how many shells there are. Write the number. Draw two more shells. Complete the addition sentence.

HOME ACTIVITY • Draw objects in a column beginning with a set of 1 to a set of 8. Have your child draw two more objects beside each set, and write how many in all.

Name _____

Add on a Ten Frame

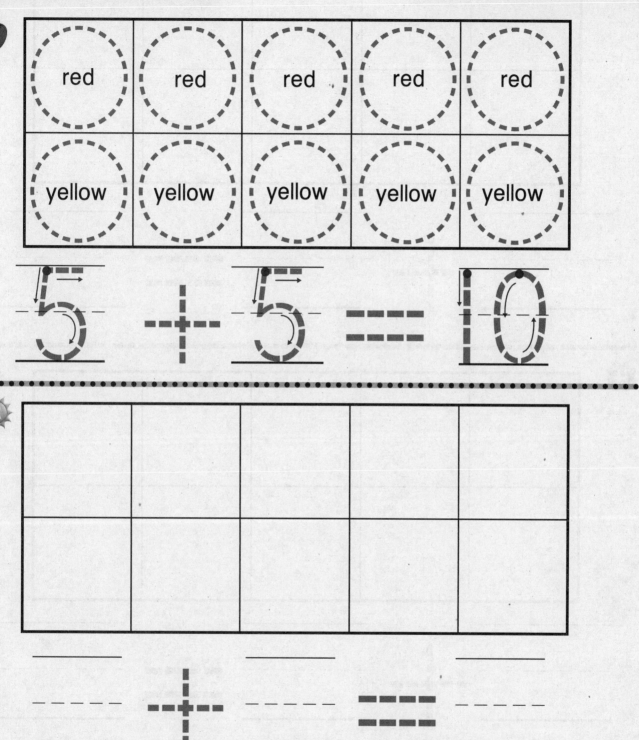

- -

DIRECTIONS 1. Place counters on the ten frame as shown.
Trace the addition sentence. 2. Place some counters red side up
on the ten frame. Add more counters yellow side up to fill the ten
frame. Complete the addition sentence.

Getting Ready for Grade 1   two hundred fifty-three **P253**

3

_____ _____ _____

_ _ _ _ $+$ _ _ _ _ $=$ _ _ _ _

_____ _____ _____

4

_____ _____ _____

_ _ _ _ $+$ _ _ _ _ $=$ _ _ _ _

_____ _____

DIRECTIONS   3–4. Place a different number of counters red side up on the ten frame. Add more counters yellow side up to fill the ten frame. Complete the addition sentence.

HOME ACTIVITY • Give your child some household objects, such as two different kinds of buttons. Have your child arrange the buttons to show different ways to make 10, such as 6 red buttons and 4 blue buttons. Write the addition sentence.

Part-Part-Whole

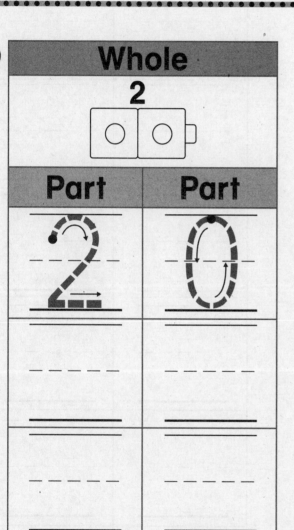

1

Whole
2

Part	Part
2	0

 2

Whole
3

Part	Part

DIRECTIONS 1–2. How many cubes are there in all? Place that many cubes in the workspace. Show the parts that make the whole. Complete the chart to show all the parts that make the whole.

Whole
4

Part	Part
_____	_____
- - - -	- - - -
_____	_____
_____	_____
- - - -	- - - -
_____	_____
_____	_____
- - - -	- - - -
_____	_____
_____	_____
- - - -	- - - -
_____	_____
_____	_____
- - - -	- - - -
_____	_____

Whole
5

Part	Part
_____	_____
- - - -	- - - -
_____	_____
_____	_____
- - - -	- - - -
_____	_____
_____	_____
- - - -	- - - -
_____	_____
_____	_____
- - - -	- - - -
_____	_____
_____	_____
- - - -	- - - -
_____	_____
_____	_____
- - - -	- - - -
_____	_____

DIRECTIONS 3–4. How many cubes are there in all? Complete the chart to show all the parts that make the whole.

HOME ACTIVITY • Have your child use buttons or macaroni pieces to show the different parts that make the whole set of 8 (e.g. 7 and 1, 6 and 2, 5 and 3, 4 and 4.)

Name _____

Equal Sets

1

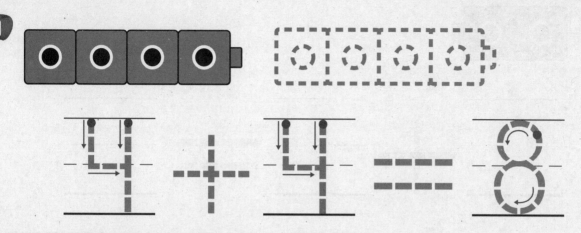

2

3

DIRECTIONS Count the cubes. Use cubes to make an equal set.
1. Trace the cubes. Trace the addition sentence. **2–3.** Draw the
cubes. Write and trace to complete the addition sentence.

4

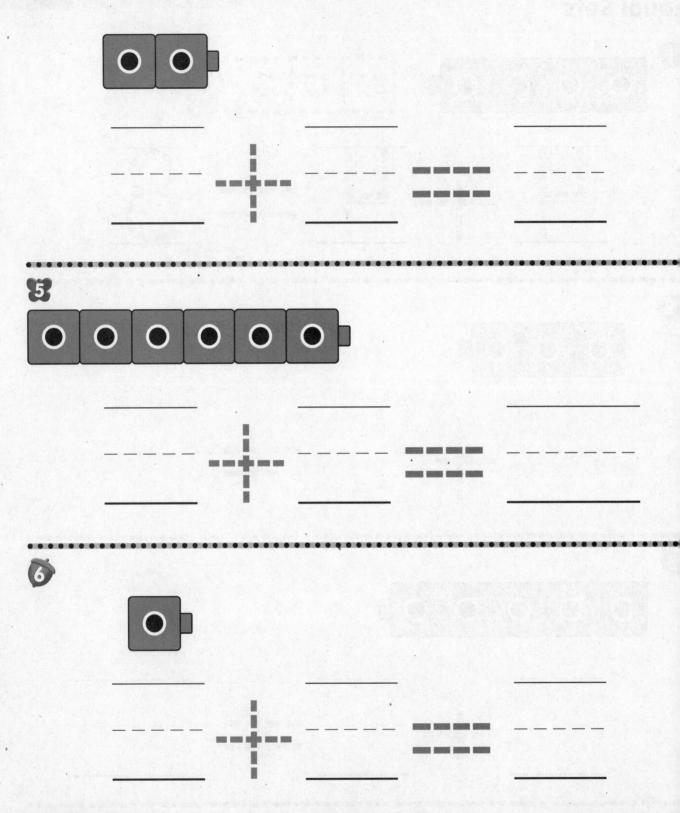

_____ _____ _____

- - - - - + - - - - - = - - - - -

_____ _____ _____

5

_____ _____ _____

- - - - - + - - - - - = - - - - -

_____ _____ _____

6

_____ _____ _____

- - - - - + - - - - - = - - - - -

_____ _____ _____

DIRECTIONS 4–6. Count the cubes. Use cubes to make an equal set. Draw the cubes. Write and trace to complete the addition sentence.

HOME ACTIVITY • Have your child show equal sets by holding up an equal number of fingers on each hand. Then have your child say the addition sentence.

Subtract to Compare

1

2 **more** ⚓

2

_ _ _ _ _

_____ **more** 👓

3

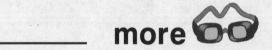

_ _ _ _ _

_____ **more**

DIRECTIONS   1. Trace the lines to match the objects in the top row to the objects in the bottom row. Compare the sets. Trace the circle that shows the set with more objects. Trace the number.   **2–3.** Draw lines to match the objects in the top row to the objects in the bottom row. Compare the sets. Circle the set that has more objects. Write how many more.

4

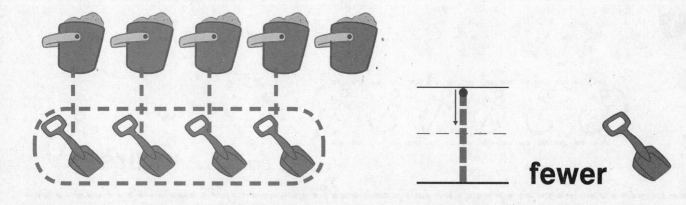

fewer

5

_ _ _ _ _ _

_ _ _ _ _ _

_ _ _ _ _ fewer

6

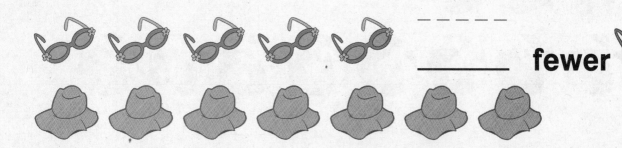

_ _ _ _ _ _

_ _ _ _ _ _

_ _ _ _ _ fewer

DIRECTIONS 4. Trace the lines to match the objects in the top row to the objects in the bottom row. Compare the sets. Trace the circle that shows the set with fewer objects. Trace the number. **5–6.** Draw lines to match the objects in the top row to the objects in the bottom row. Compare the sets. Circle the set that has fewer objects. Write how many fewer.

HOME ACTIVITY • Show your child a row of seven pennies and a row of three nickels. Have your child compare the sets, identify which has fewer coins, and tell how many fewer. Repeat with other sets of coins up to ten.

Concepts and Skills

2 ___ — ___ == ___

- -

②

___ ▬▬▬ ___ ▬▬▬▬ ___

- -

DIRECTIONS 1. Use cubes to show the number. Draw the cubes. Take away one cube. Circle the cube that you took away and mark an X on it. Complete the subtraction sentence. **(P263–P264) 2.** Place 10 counters on the ten frame. Draw the counters. Take away some counters. Circle and mark an X on the counters that you took away. Complete the subtraction sentence. **(P267–P268)**

3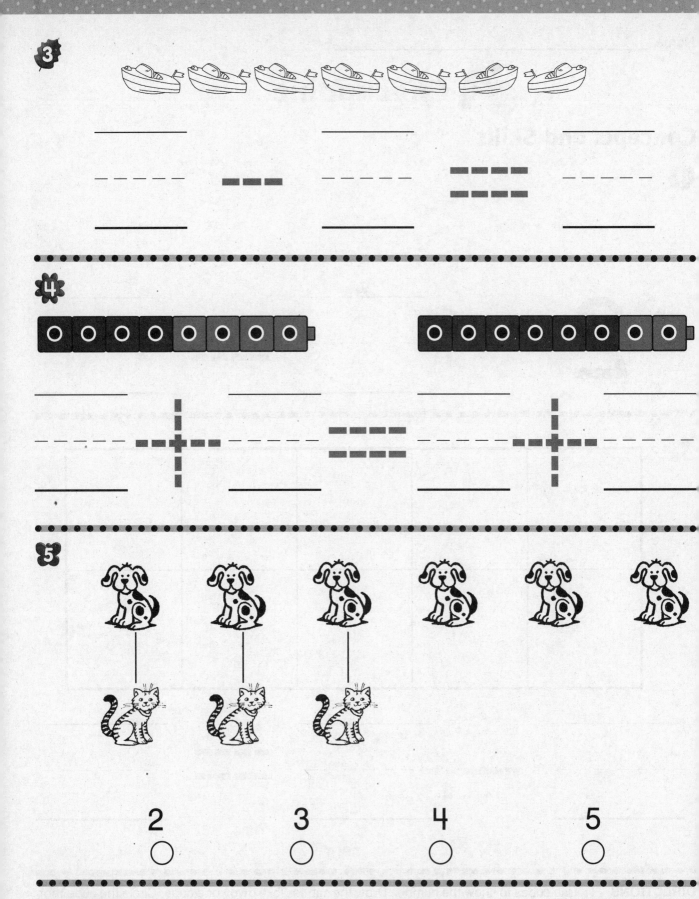

_____

– – – ▬▬▬   – – – ▬▬▬   – – – –
                               ▬▬▬
_____   _____   _____

4

_____ ╋ _____   ▬▬▬   _____ ╋ _____
                      ▬▬▬
_____   _____   _____

5

| 2 | 3 | 4 | 5 |
| ○ | ○ | ○ | ○ |

DIRECTIONS **3.** Count and write how many boats in all. Two boats leave. Circle and mark an X on those boats. Complete the subtraction sentence. **(pp. P265–P266)** **4.** Look at the cube trains. Trace and write to complete the equation. **(pp. P261–P262)** **5.** Compare the sets. Mark under the number that shows how many more dogs are shown in the picture. **(pp. P275–P276)**

Name _____

Hands On: How Many Ones?

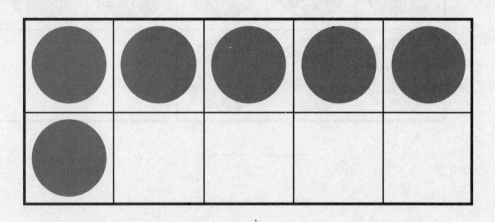

_____
- - - - - - -
_____ **ones**

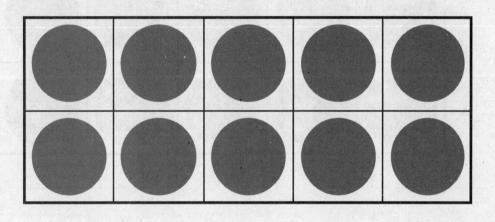

_____   _____
- - - - - - -   - - - - - - -
_____ **ones or** _____ **ten**

DIRECTIONS Place counters on the ones shown. **1.** How many ones are there? Write the number. **2.** How many ones are there? Write the number. How many tens is that? Write the number.

3

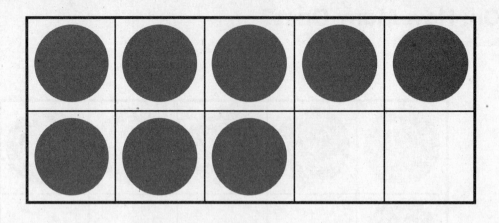

———————

— — — — —

_____ **ones**

4

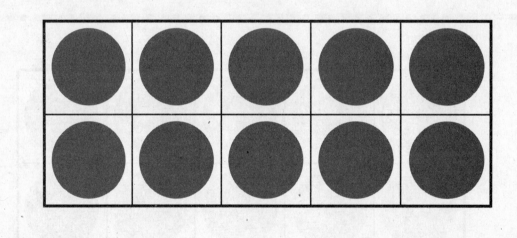

_____   _____

— — — —   — — — —

_____ **ones   or**   _____ **ten**

DIRECTIONS   Place counters on the ones shown.   **3.** How many ones are there? Write the number.   **4.** How many ones are there? How many tens is that? Write the number.

HOME ACTIVITY • Place 10 small items on a table. Ask your child to count and write how many ones that is. Then ask him or her to write how many tens that is.

Read and Write Numbers 20 to 30

  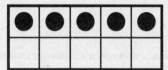 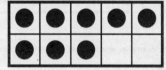

_____
- - - - - - - - - -
_____

_____
- - - - - - - - - -
_____

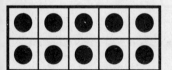 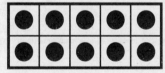 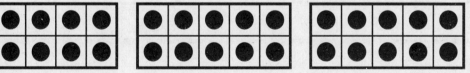

- - - - - - - - - -
_____

_____
- - - - - - - - - -
_____

DIRECTIONS How many counters are there? I. Trace the number.
2–5. Write the number.

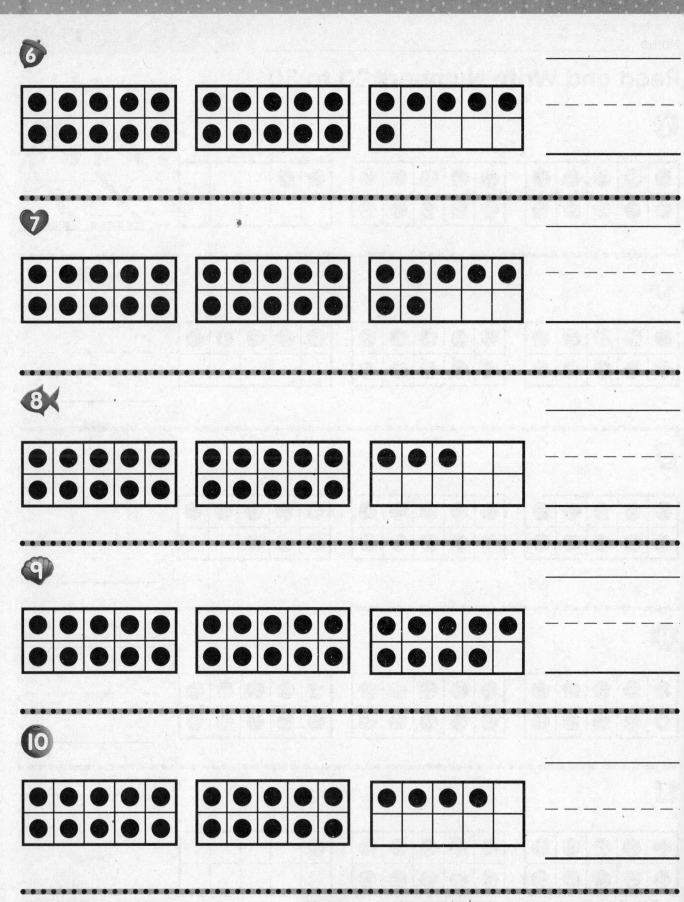

HOME ACTIVITY • Give your child 20 to 30 paper clips. Have your child count the paper clips and write how many.

Read and Write Numbers 30 to 40

1

2

3

4

5

DIRECTIONS How many counters are there? **1.** Trace the number. **2–5.** Write the number.

Getting Ready for Grade 1   two hundred eighty-three **P283**

6

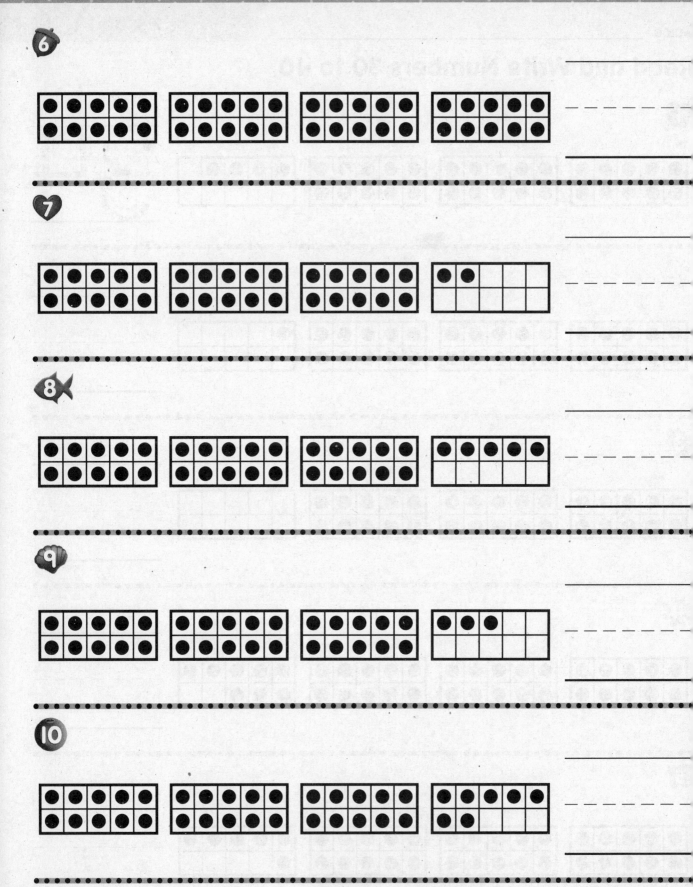

7

8

9

10

DIRECTIONS 6–10. How many counters are there? Write the number.

HOME ACTIVITY · Have your child count out cereal pieces for different numbers from 30 to 40.

Name _____

Read and Write Numbers 40 to 50

 1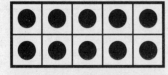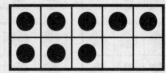

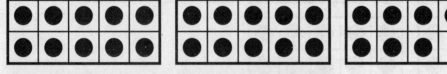 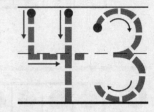

 2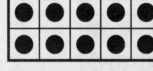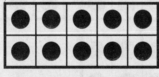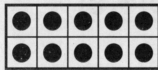

_____

- - - - - - - -

_____

 3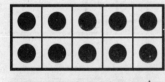

_____

- - - - - - - -

_____

 4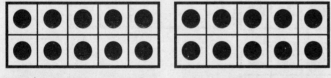

_____

- - - - - - - -

_____

DIRECTIONS How many counters are there? 1. Trace the number.
2–4. Write the number.

Getting Ready for Grade 1   two hundred eighty-five   **P285**

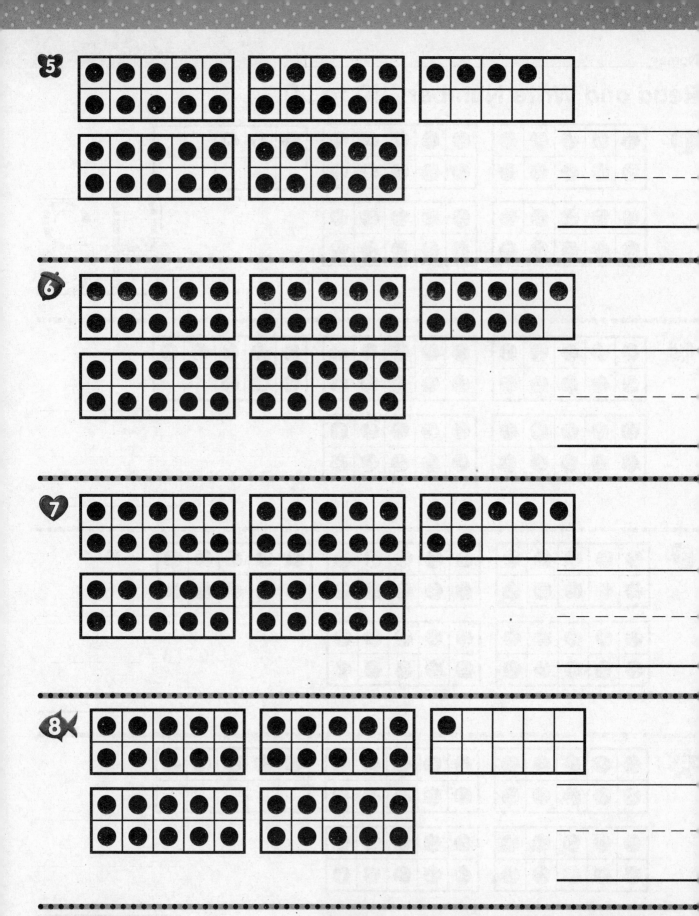

5

6

7

8

DIRECTIONS 5–8. How many counters are there? Write the number.

HOME ACTIVITY • Help your child count four sets of ten cereal pieces each. Then have him or her tell how many cereal pieces there are.

Name _____

 1

_____

_____ ones

 2

_____

- - - - - - -

 3

_____

- - - - - - -

DIRECTIONS 1. How many ones are there? Write the number.   2–3. How many counters are there? Write the number.

4.

_____
- - - - - - - - - - -
_____

5.

_____
- - - - - - - - - - -
_____

6.

_____
- - - - - - - - - - -
_____

7.

25 ◯　　30 ◯　　35 ◯　　40 ◯

DIRECTIONS 4–6. How many counters are there? Write the number.
7. How many counters are shown? Mark under the number of counters.

Numbers on a Clock

DIRECTIONS I. Trace 12 at the top of the clock. Write the numbers I to 6 in order on the clock.

Getting Ready for Grade I   two hundred eighty-nine **P289**

DIRECTIONS   2. Find 6 on the the clock. Write the numbers 7 to 12 in order on the clock.

HOME ACTIVITY · Have your child point to and name the numbers on an analog clock.

Use an Analog Clock

   o'clock

  _____ o'clock

  _____ o'clock

  _____ o'clock

DIRECTIONS 1. About what time does the clock show?
Trace the number. **2–4.** About what time does the clock show?
Write the number.

before 6 o'clock   about 6 o'clock   after 6 o'clock

 5

before 2 o'clock

about 2 o'clock

after 2 o'clock

6

before 7 o'clock

about 7 o'clock

after 7 o'clock

7

before 11 o'clock

about 11 o'clock

after 11 o'clock

DIRECTIONS 5–7. Circle the time shown on the clock.

HOME ACTIVITY • Look at or draw a simple clock. Ask your child questions such as: *Where does the hour hand go to show about 8 o'clock? About 1 o'clock? About 4 o'clock?*

Name _____

Use a Digital Clock

1

 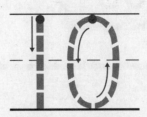 o'clock

2

_____ o'clock

3

_____ o'clock

4

_____ o'clock

DIRECTIONS   1. Trace the hour number on the digital clock. Trace to show another way to write that time.   **2–4.** Trace the hour number on the digital clock. Show another way to write that time.

6:00

_____

- - - - - - - - -

_____ **o'clock**

2:00

_____

- - - - - - - - -

_____ **o'clock**

11:00

_____

- - - - - - - - -

_____ **o'clock**

8:00

_____

- - - - - - - - -

_____ **o'clock**

DIRECTIONS   5–8. Trace the hour number on the digital clock. Show another way to write that time.

HOME ACTIVITY • Ask your child to explain or draw what a digital clock looks like at 3:00.

Checkpoint

before 9 o'clock

about 9 o'clock

after 9 o'clock

DIRECTIONS **1.** Write the missing numbers on the clock. (pp. P289–P290)
2. Circle the time shown on the clock. (pp. P291–P292)

3

7:00

——————
- - - - - - -
—————— o'clock

4

5

2   6   7   8
○   ○   ○   ○

DIRECTIONS **3.** Trace the hour number on the clock. Show another way to write that time. **(pp. P293–P294)** **4.** Write the missing numbers on the clock. **(pp. P289–P290) 5.** Mark under the number that shows about what time is on the clock. **(pp. P291-P292)**

Name _____

Use Nonstandard Units
to Express Length

 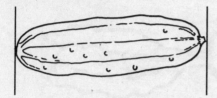

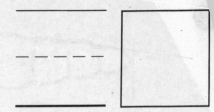

 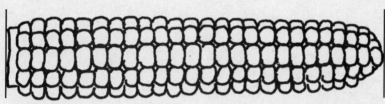

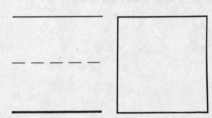

 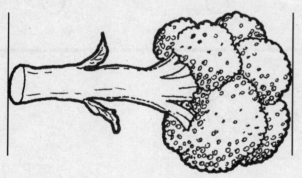

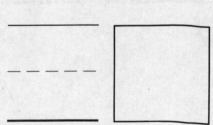

DIRECTIONS   1–3. Use squares end to end to measure how long the
vegetable is. Draw the squares. Write about how many squares long it is.

Check

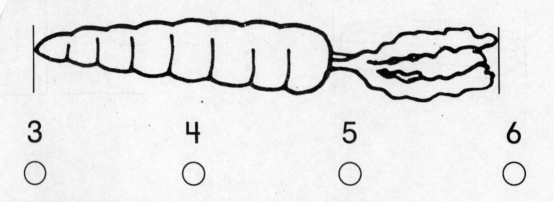

3	4	5	6
○	○	○	○

Spiral Review

7

6	8	9	10
○	○	○	○

3

1	2	3	4
○	○	○	○

DIRECTIONS   **1.** About how many squares long is the carrot? Mark under your answer. (Lesson 11.3)   **2.** Which number is less than 7. (Lesson 4.7)
3. How many birds are there? Mark under your answer. (Lesson 1.4)